VOLUME ONE HUNDRED AND SEVENTEEN

ADVANCES IN
COMPUTERS

The Digital Twin Paradigm for
Smarter Systems and Environments:
The Industry Use Cases

VOLUME ONE HUNDRED AND SEVENTEEN

ADVANCES IN
COMPUTERS

The Digital Twin Paradigm for
Smarter Systems and Environments:
The Industry Use Cases

Edited by

PETHURU RAJ
*Reliance Jio Infocomm. Ltd. (RJIL),
Bangalore, India*

PREETHA EVANGELINE
*Karunya Institute of Technology and Sciences,
Coimbatore, Tamil Nadu, India*

ACADEMIC PRESS

An imprint of Elsevier

ELSEVIER

Academic Press is an imprint of Elsevier
50 Hampshire Street, 5th Floor, Cambridge, MA 02139, United States
525 B Street, Suite 1650, San Diego, CA 92101, United States
The Boulevard, Langford Lane, Kidlington, Oxford OX5 1GB, United Kingdom
125 London Wall, London, EC2Y 5AS, United Kingdom

First edition 2020

Notices
Knowledge and best practice in this field are constantly changing. As new research and experience
broaden our understanding, changes in research methods, professional practices, or medical
treatment may become necessary.

Practitioners and researchers must always rely on their own experience and knowledge in evaluating
and using any information, methods, compounds, or experiments described herein. In using such
information or methods they should be mindful of their own safety and the safety of others, including
parties for whom they have a professional responsibility.

To the fullest extent of the law, neither the Publisher nor the authors, contributors, or editors, assume
any liability for any injury and/or damage to persons or property as a matter of products liability,
negligence or otherwise, or from any use or operation of any methods, products, instructions, or ideas
contained in the material herein.

ISBN: 978-0-12-818756-2
ISSN: 0065-2458

For information on all Academic Press publications
visit our website at https://www.elsevier.com/books-and-journals

Publisher: Zoe Kruze
Acquisition Editor: Zoe Kruze
Editorial Project Manager: Peter Llewellyn
Production Project Manager: James Selvam
Cover Designer: Alan Studholme

Typeset by SPi Global, India

Working together
to grow libraries in
developing countries

www.elsevier.com • www.bookaid.org

Contents

13. Machine learning and deep learning algorithms on the Industrial Internet of Things (IIoT) 321

P. Ambika

14. Energy-efficient edge based real-time healthcare support system

Preface

There is a growing importance of digital twins (DTs) in the increasingly connected and cognitive era. With the unprecedented adoption and adaptation of digitization, edge, and miniaturization technologies, there is a faster, risk-free, and resilient instrumentation and implementation of digitized entities in massive quantities. Digital elements are drawing unprecedented attention these days as they are innately and intelligently empowered to join in the mainstream computing. This transition ultimately contributes immensely for the production and sustenance of context-aware, mission-critical and people-centric IT systems, solutions, and services. The distinguished contributions of digital objects are projected to be trendsetting in the long run. On the other hand, with a bevy of mesmerizing advancements in the connectivity, communication, integration, and orchestration domains, we are all set to be succulently surrounded by scores of connected devices in our daily walks and works. Thus, the digitization aspect is being recognized as the crucial game changer for the total society. Precisely speaking, all our physical, mechanical, electrical, and electronics systems in our everyday environments (homes, hotels, hospitals, etc.) are technologically enabled to be digitized in order to dynamically and deftly design and deliver sophisticated functionalities, features, and facilities. Their usual operations and outputs are going to be entirely and elegantly different in the forthcoming digital era. In short, we can safely expect several noteworthy implications out of these digitization technologies and tools, which are gathering more market and mind shares nowadays.

The next-generation Internet is to comprise such kinds of digitized elements and connected devices. And their purposeful interactions are also expected to generate enormous amount of value-adding data (state, operational, analytical, transactional, performance, security, health condition, metrics, traces, logs, and environment). Therefore, the approaching Internet is being termed as the Internet of Things (IoT). Besides that, physical systems are closely integrated with cyber systems and hence we often come across the term "cyber physical systems (CPSs)."

With a greater understanding of the strategic significance of the digitization movement, we are heading toward the second breakthrough trend, which is nonetheless the faster emergence and evolution of digitalization technologies. Clouds are being positioned and perceived as the highly optimized and organized information and communication technologies (ICT) infrastructure for the ensuing digital era. In addition to the cloud infrastructure modules, there are several integrated and insightful platforms for data virtualization, ingestion, storage, analytics, mining, and visualization tasks. These platforms are being pronounced as the most suitable ones to run on cloud server machines, networking solutions, and storage appliances. Cloud solutions are famous for the cool convergence of mainframe and modern computing paradigms. Cloud infrastructures can be easily manipulated to supply the key nonfunctional requirements such as elasticity, availability, configurability, and composability. Thus, cloud applications benefit greatly. The above-mentioned platforms on cloud infrastructure can collect, cleanse, and crunch digital data in order to extricate useful and usable knowledge, which can be disseminated to actuation systems, business executives, system operators, etc., in time to empower them to ponder about the next and best course of actions with all the clarity and confidence. The data analytics scenario is seemingly bright as there are machine and deep learning algorithms emerging in order to emit out predictive and prescriptive insights. There are other artificial intelligence (AI) technologies such as computer vision, natural language processing (NLP), artificial neural networks (ANNs), etc. Clouds in consonance with data engineering, science, management, and analytics platforms are being touted as the subtle and stimulating foundation for the digital era. The digital innovation, disruption, and transformation goals are to be met fully through digital connectivity and intelligence competencies.

We have physical, networked, and embedded electronics in large numbers in our personal, social, and professional locations. It is estimated that there will be billions of connected devices in the earth planet in the years to come as per the market analysts and watchers. But the point here is that some of them are very complicated yet sophisticated to design, develop, test, debug, and operate. And also, they need extra and continuous attention due to their mission-critical contributions. For example, specialized engines at manufacturing floors, medical instruments, humanoid robots assisting surgeons, drones delivering packages and parcels, satellites and their launchers under development, train and aircraft engines, submarines, etc., are the dominant and prominent devices yearning for pioneering technologies and tools to be beneficial for their users. Automotive electronics, communication gateways, avionics, etc., are to get immeasurably benefited through constant monitoring, measurement, and management.

But there are significant challenges. By the way, their development and operational complexities are unusually on the higher side. Their functioning, deliveries, and interactions are very much unfathomable and unpredictable. The challenge and concern here are to moderate their complexities substantially, proactively pinpoint their risks and rewards, clearly understand their resiliency and reliability attributes, sharply enhance their intelligence through data-driven insights, critically analyze and predict their performance, health condition, etc., confidently take insights driven decisions and actions, etc. It can be a single device or a cluster of devices at the ground. Thus, there came a few interesting solution approaches to solve these limitations in the increasingly connected era. The successful and sustainable idea here is to create and come out with a virtual representation for such devices under keen observation.

Digital twins (DTs) are the exact virtual/logical/cyber/digital representations/replicas of ground-level physical devices. With the surging popularity of cloud environments, the DTs of electronics, machines, equipment, instruments, appliances, wares, utilities, drones, robots, and other worthwhile electronics are being produced and deposited in cloud storages. Because of the consistent and continuous communication with corresponding real-world and physical devices, the digital version of the physical device comes handy in fulfilling emerging and evolving requirements. Data lakes put up in large-scale cloud environments, the faster stability and maturity of AI algorithms and approaches, the ready availability of data analytics platforms for performing bulky and batch processing besides streaming analytics, the real-time and runtime data being emitted by ground-level physical twins to be shared across to their respective digital twins and other developments foretell that there are bright days in producing intelligent devices. Before producing any physical device, we can easily understand its characteristics through its DT. Thus, all kinds of possibilities, opportunities, associations, patterns, and other actionable insights can be generated and used before the physical device is built and deployed. Thus, DTs bring forth a number of tactical as well as strategically sound advantages. Several business domains are embracing this irreversible and irresistible paradigm. The prominent application domains include smart cities, smart manufacturing, etc. The DT phenomenon is bound to be a key contributor for the mission and vision of industry 4.0.

This book is specially crafted with the ultimate aim of conveying the current trends and coordinated efforts being undertaken to take this method more promising, pervasive, and persuasive. There are 14 chapters as enunciated below:

1. Demystifying the digital twin paradigm—This is to give a detailed overview of this new technological paradigm. The humble beginning, the current happenings, and the future direction are described in this chapter for the benefit of our esteemed readers.

2. Digital twin technology for "smarter manufacturing"—As articulated above, smart manufacturing is one of the critical application areas of the DT aspect. This chapter has explained how the flourishing DT idea is to support and sustain smart manufacturing.

3. The fog computing/edge computing to leverage digital twin— Nowadays the data processing happens where the data sources are. How the combination of edge computing and digital twin is going to be a trend-setting affair in the days to come is vividly delineated in this chapter.

4. The industry use cases for the digital twin idea—There are several industry domains embracing this new phenomenon in order to be ahead of their peers in this cut-throat competitive environment. This chapter has thrown a lot of information toward that.

5. Enabling digital twin at the edge—The edge computing and analytics capabilities are being insisted in order to realize real-time application development. Forming edge clouds dynamically in an ad hoc manner is one such approach to simplify and streamline edge computing. This chapter has talked about an edge device and how it can collaborate with its corresponding cloud-based digital twin to bring in additional intelligence.

6. The industrial internet of things (IIOT)—This chapter talks about the implications of various industrial machineries getting connected and synchronized with one another in the vicinity and with remotely held cloud applications, services, and databases. As indicated above, the value of DT for industrial machines at the ground level is bound to sharply rise.

7. The growing role of integrated and insightful big and real-time data analytics platforms—Data analytics is an important requirement for the intended success of the DT concept. There are big, fast, and streaming data emanating from all kinds of IoT, IIoT, and cyber physical systems. Herein, data analytics platforms work with DT to extricate useful and usable information. There are systems helping out in real-time analytics of big data.

8. Air pollution control model using machine learning and IoT—This chapter describes about a specific use case leveraging the ML, IoT, and DT.

9. The human body: A digital twin of cyber physical systems—This chapter has a unique value proposition. How a human body is a digital twin for cyber physical systems is clearly put forth in this chapter.

10. Impact of cloud security in digital twin—The security, as usual, is a huge challenge in the digital era. This chapter throws more light on the impacts of cloud security on DTs.

11. Digital twin in consumer choice modeling—This is another interesting use case of digital twin. The consumer choice modeling can be elegantly simplified through the application of the advancements happening in the DT space.

12. Digital twin: The industry use cases—This chapter lists out a series of powerful and prominent use cases of the DT paradigm.

13. Machine and deep learning algorithms on the Industrial Internet of Things (IIOT)—As enunciated above, machine and deep learning (ML/DL) algorithms are very essential for the strategic success of the DT technology. This chapter exclusively focuses on demystifying the various ML and DL algorithms that can be used to predict and prescribe a variety of things in consonance with DT.

14. Energy efficient edge based real-time healthcare support system—This is again an edge use case. The DT will play a very crucial role in shaping up edge devices to be right and relevant for their users.

CHAPTER ONE

Stepping into the digitally instrumented and interconnected era

Pethuru Raj[a], Jenn-Wei Lin[b]
[a]Reliance Jio Infocomm. Ltd. (RJIL), Bangalore, India
[b]Department of Computer Science and Information Engineering, Fu Jen Catholic University, New Taipei City, Taiwan

Contents

Advances in Computers, Volume 117
ISSN 0065-2458
https://doi.org/10.1016/bs.adcom.2019.09.008

Abstract

This chapter is to tell all about the digitization-inspired possibilities and opportunities and how software-defined cloud centers are the best fit for hosting and running digital applications. Also, how the next-generation data analytics can be smartly accomplished through cloud platforms and infrastructures is also explained in detail. We are to describe some of the impactful developments and technological advancements brewing in the IT space, how the tremendous amount of data getting produced and processed through cloud systems is to impact the IT and business domains, and how next-generation IT infrastructures are accordingly getting refactored, remedied and readied for the impending big data-induced challenges, how likely the move of the data analytics discipline toward fulfilling the digital universe requirements of extracting and extrapolating actionable insights for the knowledge-parched is, and finally for the establishment and sustenance of the dreamt smarter planet. In short, the uninhibited explosion of digitized systems and connected devices pour out a tremendous amount of multi-structured data and the impending challenge is to make sense out of the data heaps. Data analytics is the way to go and in the recent past, the overwhelming trend is to empower our everyday systems with machine and deep learning algorithms to automatically learn out of data heaps and streams in order to be distinctively intelligent in their actions and reactions. This chapter is specially prepared to put a stimulating foundation for explaining the nitty-gritty of the Digital Twin paradigm.

1. Introduction

There is no doubt that the aspects of digitization and digitalization are buzzing across industry verticals these days. The digitization is all about empowering commonly found and tangible things in our midst to be digitized entities by systematically applying proven and potential edge technologies. There is a renewed interest by researchers and professionals toward unearthing scores of pioneering edge technologies such as bar codes, microchips, microcontrollers, miniaturized stickers, disappearing labels and RFID tags, networking sensors and actuators and beacons, LED lights, etc. The faster maturity and stability of the edge technologies speeds up the production and deployment of powerful digitized objects, alternatively termed as smart or sentient materials. These digitized elements are self-, surroundings- and situation–aware. These empowered systems subsequently work and contribute toward the mainstream computing. These, being small in size and large in number, can be readily attached and implanted with any physical, mechanical and electrical systems in our daily environments to make them ready to be digitized and connected. Thus, empowering all kinds of concrete products and tangible things gains the speed. The enigmatic embedded space is bound to grow tremendously bringing forth hitherto unheard possibilities and opportunities for the human society.

That is, we are being bombarded with sensitive, perceptive, and responsive artifacts in and around us. Next in line is that the explosion of instrumented and connected devices that may be purpose-generic as well as specific. There are portable, wearable, wireless, mobile, implantable, hearable and nomadic devices in plenty. The device ecosystem is growing rapidly. We have sophisticated, slim and sleek, trendy and handy devices assisting humans in their everyday tasks. Market research analysts and watchers have forecast that there will be billions of connected devices and trillions of digitized entities in the years ahead. The flourishing domains of the Internet of things (IoT) and cyber-physical systems (CPS) are predominantly representing the digitization movement. The implications are many and diverse. We hear and read about smart manufacturing, cities, hotels, homes, retail, healthcare, defense, logistics, etc. Precisely speaking, the networked embedded systems are to lead to smarter and sophisticated systems, networks, services and environments.

With digitization sweeping across the industry verticals, we have a greater number of connected devices. When they interact with one another, there results a large amount of multi-structured data. It becomes mandatory to collect and make sense out of that data heaps. Now we talk about digitalization. There are several digital technologies such as software-defined cloud environments, integrated data analytics platforms, enterprise mobility, security, etc. That is, the cloud idea represents the IT infrastructure optimization and organization. Newer architectural patterns and styles (microservices architecture (MSA) and event-driven architecture (EDA)) are emerging and evolving fast for designing and building enterprise-grade applications. Integrated analytics platforms for collecting, cleansing, preprocessing, storing, analyzing and visualizing insights out of big, fast, and streaming data are made available in plenty. The tremendous improvements in machine and deep learning algorithms come handy in doing advanced and automated analytics of transactional, interaction, analytical, operational and business data in order to extract personalized, predictive and prescriptive insights. The era of mobile apps is literally mesmerizing the world. Further on, enterprise, cloud and web applications are empowered to have mobile interfaces. With all the open data and the increasingly connected world, security is the primary concern for citizens. Thus, the discipline of cyber security has drawn a lot of attention of security experts and researchers. There are security–related algorithms, approaches and architectures in order to ensure tighter and impenetrable security for data (at rest, transit and usage), digital applications, network and infrastructures.

Thus, highly optimized IT infrastructures, integrated analytics platforms, microservices architecture (MSA), security appliances, artificial intelligence (AI) advancements, enterprise mobility, etc., are being collectively leveraged to ring in real digital transformation. The faster proliferation of ground-level digitized/IoT elements and their seamless and spontaneous integration with cyber/virtual applications being run on multiple cloud infrastructures is to succulently strengthen the digital era.

There are several strategically sound implications of these unique transitions for not only businesses but also for every individual, innovator and institution. Digital innovation, disruption and intelligence are some of the top buzzwords in the industry arena with the widespread adoption of digitization and digitalization technologies and tools. The widely discussed and discoursed digital era is beckoning and reckoning us with the faster stability and maturity of potential and promising digital technologies. The prospect of digital transformation is definitely blooming with the greater awareness among business executives and IT professionals. An arsenal of digital transformation-enabling processes, platforms, patterns, and practices are being unearthed in order to simplify and streamline the activities that eventually lead to the promised digital era. There is a growing family of pioneering technologies, and tools emerging and evolving to swiftly fulfill up the digitalization goals. These spectacular turnarounds bring forth a kind of beneficial convergence and in that process, the physical and virtual worlds are being systematically integrated.

On summarizing, all kinds of physical, mechanical, electrical, and electronics systems are adequately empowered to be instrumented and interconnected to exhibit intelligent behavior. Further on, all kinds of connected devices and digitized objects in our everyday environments are systematically integrated with remotely held and cloud-hosted applications, services, and data sources in order to be adaptive in their deeds, decisions, and deliveries.

2. Elucidating digitization technologies

One of the most visible and value-adding trends in IT is nonetheless the digitization aspect. All kinds of everyday items in our personal, professional and social environments are being digitized systematically to be computational, communicative, sensitive, perceptive, and responsive. That is, all kinds of ordinary entities in our midst are technologically instrumented and interconnected to be extraordinary in their operations, outputs, and offerings. These days, due to the unprecedented maturity and stability of a host of path-breaking technologies such as miniaturization, integration, communication, computing, sensing, perception, middleware, analysis, actuation and orchestration, everything in our environments has grasped the inherent power of finding and binding with one another in its vicinity as well as with remote objects via networks purposefully and on need basis to uninhibitedly share their distinct capabilities toward the goal of business automation, acceleration and augmentation. Ultimately, every small and tangible thing will become smart, electronics goods will become smarter and human beings will become the smartest in their deals, decisions, and deeds. In this section, the most prevalent and pioneering trends and transitions in the IT landscape will be discussed. Especially the digitization technologies and techniques are given the sufficient thrust.

2.1 The trend-setting technologies in the IT space

As widely reported, there are several delectable transitions in the IT landscape. The consequences are definitely vast and varied: the incorporation of nimbler and next-generation features and functionalities into existing IT solutions and the eruption of altogether new IT products and solutions for the humanity. These have the intrinsic capabilities to bring forth numerous subtle and succinct transformations in business as well as people. Businesses are being empowered with newer possibilities and opportunities.

2.2 IT consumerization

There are much-discoursed and deliberated reports detailing the diversity and availability of mobile devices (smartphones, tablets, wearables, drones, robots, etc.) and their management platforms. There are mobile apps in plenty. This trend is ultimately empowering people in their daily works and walks. The ubiquitous information access is made possible. Further on, the IT infrastructures are being tweaked accordingly in order to gracefully support this strategically sound movement. There are some challenges for IT administrators in fulfilling the device explosion. That is, IT is steadily becoming an inescapable part of consumers directly and indirectly. And the need for robust and resilient device management software with the powerful emergence of "bring your own device (BYOD)" paradigm is being felt and is being insisted across. Another aspect is the emergence of next-generation mobile applications across a variety of business verticals. There is a myriad of development platforms, programming and markup languages, enabling frameworks, tools, and lightweight operating systems in the fast-moving mobile space. Precisely speaking, IT is not only for businesses but also for every human being.

2.3 IT commoditization

This is another cool trend penetrating in the IT industry. With the huge acceptance and adoption of cloud computing and big data analytics, the value of commodity IT is decidedly on the rise. The embedded intelligence inside IT hardware elements is being abstracted and centralized through hypervisor software solutions. Hardware systems are thus software-enabled to be flexibly manipulated and programmed. With this transition, all the hardware resources in any data center become dumb and can be easily replaced, substituted, and composed for easily and quickly fulfilling different requirements and use cases. Commoditized IT solutions are relative cheap and hence the goal of the IT affordability is thus realized along with a number of other advantages. That is, the future IT data centers and server farms are going to be stuffed with a number of commodity servers, storages, and network solutions.

2.4 IT compartmentalization (virtualization and containerization)

The "divide and conquer" method has been the most versatile and rewarding mantra in the IT field. Abstraction is another powerful and established technique in the IT space. The widely used virtualization, which had laid a stimulating and sustainable foundation for the raging cloud idea, is actually hardware virtualization. The virtualization has penetrated into storage appliances, network components and security solutions. That is, entire data centers are methodically virtualized. There are a few serious drawbacks with virtualization.

Then the aspect of containerization being represented through the popular Docker platform. Containerization is the operating system (OS)-level virtualization. Containers are lightweight and hence attain the native performance of the physical machines. The real-time horizontal scalability is being facilitated by the concept of containerization. These two are clearly and cleverly leading to the cloud era.

The cloud idea represents the *IT industrialization*. Consolidating, virtualizing and/or containerizing and centralizing all kinds of IT systems, putting it on-premise and/or off-premise, operating them in a shared and automated fashion, delivering them in an online and on-demand manner, continuously adding newer capabilities by bringing forth fresh technologies and tools, etc., lead toward industrialized IT.

2.5 IT digitization and distribution

As explained in the beginning, digitization has been an on-going process and it has quickly generated and garnered a lot of market and mind shares. Digitally enabling everything around us induces a dazzling array of cascading and captivating effects in the form of cognitive and comprehensive transformations for businesses as well as people. With the growing maturity and affordability of scores of edge technologies, every common thing in our personal, social, and professional environment is becoming digitized. Everyday devices are being systematically empowered to be intelligent. Ordinary articles are becoming smart artifacts in order to significantly enhance the convenience, choice, and comfort levels of humans in their everyday lives and works.

Similarly, the distribution aspect too gains more ground. Due to its significant advantages in crafting and sustaining a variety of large-scale business applications, the distributed computing phenomenon has become popular. Distributed applications, though weighed down by security implications, are good in fulfilling various non-functional requirements (NFRs) such as availability, scalability, modifiability, accessibility, etc. Lately, there is a bevy of software architectures, frameworks, patterns, practices, and platforms for realizing distributed applications.

2.6 Why digitization?

Ultimately all kinds of perceptible and concrete objects in our everyday environments will be empowered to be self-, surroundings-, and situation-aware; remotely identifiable; readable; recognizable; addressable; and controllable.

Such a profound empowerment will bring forth real transformations for the total human society, especially in establishing and sustaining smarter environments, such as smarter homes, buildings, hospitals, classrooms, offices, and cities. Suppose, for instance, a man-made or natural disaster occurs. If everything in the disaster area is digitized, then it becomes possible to rapidly determine what exactly has happened, the intensity of the disaster, and the hidden risks of the affected environment. Any information extracted provides a way to properly plan and proceed insightfully, reveals the extent of the destruction, and conveys the correct situation of the people therein. The knowledge gained would enable the rescue and emergency team leaders to cognitively contemplate appropriate decisions and plunge into actions straightaway to rescue as much as possible thereby minimizing damage and losses to properties and people.

In short, digitization will substantially enhance our decision-making capability in our personal as well as professional lives. Digitization also means that the ways we learn and teach are to change profoundly, energy usage will become knowledge-driven so that green goals can be met more smoothly and quickly, and the security and safety of noble things will go up considerably. As digitization becomes pervasive, our living, relaxing, working, and other vital places will be filled up with a variety of electronics including environment-monitoring sensors, actuators, disappearing controllers, projectors, cameras, appliances, high-definition IP TVs, and the like. In addition, items such as furniture and packages will become empowered by attaching state-of-the-art LEDs, beacons, infinitesimal sensors, specialized electronics, communication modules, etc. Whenever we walk into such kinds of enlightened environments, the devices we carry and even our e-clothes will enter into a collaboration mode to form wireless and ad hoc networks with the digitized objects in that environment. For example, if someone wants to print a document from his smartphone or tablet, and he enters into a room where a printer is installed, the smartphone will automatically begin a conversation with the printer, check its competencies, and send the documents to be printed. The smartphone will then alert the owner about the neat and nice accomplishment.

Digitization will also provide enhanced care, comfort, choice, and convenience. Next-generation healthcare services will demand deeply connected and cognitive solutions. For example, ambient assisted living (AAL) is a new prospective application domain where lonely, aged, diseased, bedridden and debilitated people living at home will receive are remote diagnosis, care, and management as medical doctors, nurses and other caregivers remotely monitor patients' health parameters.

People can track the progress of their fitness routines. Taking decisions become an easy and timely affair with the prevalence and participation of connected solutions that benefit knowledge workers immensely. All the secondary and peripheral needs will be accomplished in an unobtrusive manner so that people nonchalantly focus on their primary activities. However, there are some areas of digitization that need some attentions. One is the goal of energy efficiency. Green solutions and practices are being insisted upon everywhere these days, and IT is one of the principal culprits in wasting a lot of precious energy due to the pervasiveness of commoditized IT servers and connected devices. Data centers armed with a large number of server machines, storage appliances and networking solutions are bound to consume a lot of electricity and they dissipates more heat to the fragile environment. So green IT has become a hot subject for the deeper and decisive study and research across the globe. Another prime area of interest is remote monitoring, management, and enhancement of the empowered devices. With the number of devices in our everyday environments is growing at an unprecedented scale, their real-time administration, configuration, activation, monitoring, management, and repair (if any problem arises) can be eased considerably with effective remote connection and correction competency.

2.7 Extreme connectivity

The connectivity trait has risen dramatically and become deeper and extreme. The network topologies and technologies are consistently expanding and empowering their participants and constituents to be highly productive. There are unified, ambient and autonomic communication technologies emanating from worldwide research organizations and labs. These transitions draw the attention of executives and decision-makers in a bigger way. All kinds of digitized elements are intrinsically empowered to form ad hoc networks for accomplishing specialized tasks in a simpler and smarter manner. There are a variety of network and security solutions in the form of load balancers, switches, routers, gateways, proxies, firewalls, etc., and these are nowadays available as hardware and software appliances.

2.8 Device middleware or device service bus (DSB)

DSB is the latest buzzword enabling a seamless and spontaneous connectivity and integration between disparate and distributed devices. That is, device-to-device (in other words, machine-to-machine (M2M)) communication is the talk of the town. The interconnectivity-facilitated interactions among diverse categories of devices precisely portend a litany of supple, smart and sophisticated applications for people. Due to the multiplicity and heterogeneity of devices, the device complexity is to rise further. Device middleware solutions are being solicited in order to substantially enable devices to talk to one another in the vicinity and with remote ones through appropriate networking.

2.9 Software-defined networking (SDN)

SDN is the latest technological trend captivating professionals to have a renewed focus on this emerging yet compelling concept. With clouds being strengthened as the core, converged and central IT infrastructure, the scenarios for device-to-cloud interactions are fast-materializing. This local, as well as remote connectivity, empowers ordinary articles to become extraordinary objects by distinctively communicative, collaborative, and cognitive.

Another associate is *network function virtualization (NFV)*. With virtualized and containerized networking facilities, devices and clouds are to find, bind and capitalize one another to achieve bigger and better things. Thus, networks are being readied through versatile technological solutions to be adaptive in order to succulently support smart systems and environments.

2.10 Service enablement

Technologies with some stuff are bound to be used widely. Businesses and IT companies embrace potential and promising technologies. For example, the Internet computing has led to the development and deployment of scores of web applications. That is, customer-facing applications got web-enabled. With the pervasiveness of slim and sleek, handy and trendy and multi-faceted smartphones, enterprise and cloud applications are being mobile-enabled. That is, many applications are being accessed through mobiles on the move. This transition fulfills the longstanding goal of ubiquitous information access. Enterprise mobility, therefore, has become an interesting trend. With the overwhelming approval of the service idea, every system, device, and application, is being systematically service-enabled. We, therefore, often read, hear and feel service-oriented systems stuffed with service APIs. The service-enablement idea has brought in several spectacular changes. With the growing device ecosystem, the service-orientation phenomenon comes as a blessing in disguise. That is, we do not see any device as a hardware entity anymore. Instead, we see everything as a service. As we all know, the communication and data transmission protocols, varied data formats and implementation technologies are heavily differing in the device space. These differences and deficiencies are being are minimized through service-enablement.

Physical devices at the ground level are being seriously service-enabled in order to uninhibitedly join in the mainstream computing tasks. That is, devices, individually and collectively, could become service providers or publishers, brokers and boosters, and consumers. The prevailing and pulsating idea is that any service-enabled device in a physical environment could interoperate with others in the vicinity as well as with remote devices and applications. Services could abstract and expose only the specific capabilities of devices through service interfaces while service implementations are hidden from user agents. Such a smart separation enables any requesting device to see only the capabilities of target devices, and then connect, access, and leverage those capabilities to achieve business or people services. The service enablement completely eliminates all dependencies so that devices could interact with one another flawlessly and flexibly.

Further on, application interoperability gets accomplished through the widely adopted service-enablement facet. Ultimately, the majority of next-generation, enterprise-scale, mission-critical, process-centric and multi-purpose applications are being assembled out of multiple discrete and complex services.

2.11 The internet of things (IoT)/the internet of everything (IoE)

Originally, the Internet was the global network of networked computers. Then, with the heightened ubiquity and utility of wireless and wired devices, the scope, size, and structure of the Internet has changed vastly. This phenomenon is being touted as the Internet of Devices (IoD). With the faster adoption of service-oriented architecture (SOA), microservices architecture (MSA) and event-driven architecture (EDA), services emerge as a viable building-block for constructing enterprise-grade applications. Services are distributed and decentralized. Picking and composing them result in business and people-centric applications. With the service paradigm being positioned as the most optimal, rational and practical way of building enterprise-class applications, a gamut of services (business and IT) are being built by many, deposited in public as well as private software repositories, deployed in web and application servers to be found and used by everyone via an increasing array of multimodal and multi-faceted input/output devices. The increased accessibility and audibility of services have propelled software architects and engineers to visualize and realize modular, scalable and secure software applications by choosing and composing appropriate services from those service repositories quickly. Thus, the Internet of services (IoS) idea is fast-growing. Another interesting phenomenon getting the wider attention of press these days is the Internet of energy. That is, our personal, as well as professional devices, get their energy through their interconnectivity. Precisely speaking, how different things are linked with one another in order to conceive, concretize and deliver futuristic services for the mankind in an energy-efficient manner.

As digitization gains more accolades and success, all sorts of everyday objects are being connected with one another as well as with scores of remote applications in cloud environments. That is, everything is becoming a data supplier for the next-generation applications thereby becoming an indispensable ingredient individually as well as collectively in consciously conceptualizing and concretizing smarter applications. There are several promising implementation technologies, standards, platforms, and tools enabling the realization of the IoT vision. The probable outputs of the IoT field is a cornucopia of smarter environments such as smarter offices, homes, hospitals, retail stores, cities, etc. Cyber-physical systems (CPS), ambient intelligence (AmI), and ubiquitous computing (UC) are some of the related concepts encompassing the ideals of IoT.

In the upcoming era, unobtrusive computers, communication modules, and invisible sensors will be facilitating decision-making in a smart way. Computers in different sizes, look, capabilities and interfaces will be fitted, glued, implanted, and inserted everywhere to be coordinative, calculative, and coherent in their actions. The interpretation and involvement of humans in operationalizing these sophisticated and sentient objects are almost zero. With autonomic IT infrastructures, more intensive and insightful automation and orchestration are bound to happen. Devices collectively will also be handling all kinds of everyday needs. Drones will be pervasive and humanized robots extensively will get used in order to fulfill our daily physical chores.

On summarizing, the Internet is fast expanding. Manufacturing machines, medical instruments, defense equipment, home appliances, everyday devices, specialized robots and drones, kitchen utensils, consumer electronics, infinitesimal sensors, etc., will be linked up with the Internet in order to get more people–centric use cases.

2.12 Tending toward the trillions of digitized elements/smart objects/sentient materials

The surging popularity of digitization and edge technologies such as implantables, wearables, hearables, portables, microcontrollers, miniaturized

RFID tags, easy-to-handle barcodes, stickers, and labels, nano-scale specks, and particles, illuminating LED lights, etc., is leading the digital transformation. These come handy in readily enabling every kind of casual and cheap things in our everyday environments to be computational, communicative, sensitive, perceptive, responsive, and active. All kinds of tangible things in our midst get succulently transitioned into digitized to perform hitherto unheard intelligent services.

Further on, every kind of physical, mechanical and electrical system in the ground level get hooked to various software applications and data sources at the faraway as well as nearby cyber/virtual environments (cloud). Resultantly, the emerging domain of cyber-physical systems (CPS) is gaining immense attention lately. It is forecast that there will be trillions of digitized articles in the years ahead. The table below lists out the prominent and dominant IoT technologies for realizing IoT devices and applications.

1. The Realization technologies are maturing (Miniaturization, Instrumentation, Connectivity, remote programmability/service-enablement/APIs, sensing, vision, perception, analysis, knowledge engineering, Decision-enablement, etc.)

2. A flurry of edge technologies (sensors, stickers, specks, smart dust, codes, chips, controllers, LEDs, tags, actuators, etc.)

3. Ultra-high bandwidth communication technologies (wired as well as wireless (4G, 5G, etc.))

4. Low-cost, power and range communication standards: LoRa, LoRaWAN, NB-IoT, 802.11 × Wi-Fi, Bluetooth Smart, ZigBee, Thread, NFC, 6LowPAN, Sigfox, Neul, etc.

5. Powerful network topologies, Internet gateways, integration and orchestration frameworks, and transport protocols (MQTT, UPnP, CoAP, XMPP, REST, OPC, etc.) for communicating data and event messages

6. A variety of IoT application enablement platforms (AEPs) with application building, deployment and delivery, data and process integration, application performance management, security, orchestration, and messaging capabilities

7. Event Processing and Streaming Engines are for event message capture, ingestion, processing, etc.

8. A bevy of IoT data analytics platforms for extracting timely and actionable insights out of IoT data

9. Edge/Fog Analytics through Edge Clouds

10. IoT Gateways, platforms, middleware solutions, databases, and applications on cloud environments

2.13 Ticking toward the billions of connected devices

A myriad of electronics, machines, instruments, wares, equipment, drones, pumps, robots, smartphones, and other devices across industry verticals are intrinsically instrumented at the design and manufacturing stages in order to embed the connectivity capability. These devices are also being integrated with software services and data sources at cloud environments to be enabled accordingly. Thus, there will be billions of connected devices in the years to unfold.

2.14 Envisioning millions of software services

With the accelerated adoption of microservices architecture (MSA), enterprise-scale applications are being expressed and exposed as a dynamic collection of fine-grained, loosely-coupled, network-accessible, publicly discoverable, API-enabled, composable and lightweight services. This arrangement is all set to lay down a stimulating and sustainable foundation for producing next-generation software applications out of distributed microservices. The emergence of the scintillating concepts such as Docker containers, container orchestration platforms, and DevOps is to lead the realization of cloud-native applications in conjunction with the MSA pattern. Not only software services but also there are hardware systems are also presented as services. That is, hardware resources are being software-defined in order to incorporate the much-needed flexibility, maneuverability, and extensibility. In short, every tangible thing becomes smart, every device becomes smarter and every human being tends to become the smartest.

The disruptions and transformations brought in by the series of delectable advancements are really mesmerizing. The IT has touched every small or big entity decisively in order to produce context-aware, service-oriented, event-driven, knowledge-filled, people-centric, and cloud-hosted applications. Data-driven insights and insights-driven enterprises are indisputably the new normal.

2.15 Infrastructure optimization

The entire IT stack has been going for the makeover periodically. Especially on the infrastructure front, due to the closed, inflexible, and monolithic nature of conventional IT infrastructures, there are concerted efforts being undertaken by many in order to untangle them into modular, open, extensible, converged, and programmable infrastructures. Another worrying factor is the underutilization of expensive IT infrastructures (servers, storages, and networking solutions). With IT becoming ubiquitous for automating most of the manual tasks in different business verticals, the problem of IT sprawl is to go up and they are mostly underutilized and sometimes even unutilized for a long time. Having understood these prickling issues pertaining to IT infrastructures, the concerned have plunged into unveiling versatile and venerable measures for enhanced utilization and infrastructure optimization. Infrastructure rationalization and simplification are related activities. That is, next-generation IT infrastructures are being realized through consolidation, centralization, federation, virtualization, containerization, automation, and sharing. To bring in more flexibility, software-defined infrastructures are being proclaimed and prescribed these days.

With the faster spread of big data analytics platforms and applications, commodity hardware is being insisted for big data storage and processing. That is, we need low-cost and power infrastructures with supercomputing capability and virtually infinite storage. The answer is that all kinds of underutilized servers are collected and clustered together to form a dynamic and huge pool of server machines to efficiently tackle the increasing and intermittent needs of computation. Precisely speaking, clouds are the highly optimized and organized infrastructures that fully comply with the evolving expectations elegantly and economically. The cloud technology, though not a new one, represents a cool and compact convergence of several proven technologies to create a spellbound impact on both business and IT. Clouds emerge as the one-stop solution for all kinds of IT requirements. The cloudification represents the virtual IT era. As there is a tighter coupling between the physical and the cyber worlds, the distinct contributions of the cloud paradigm are vast and varied.

The tried and tested technique of "divide and conquer" in software engineering is steadily percolating to hardware engineering. Decomposition of physical machines into a collection of sizable and manageable virtual machines/containers for enhanced resource utilization and on the other hand, these virtual machines can be composed to create a virtual supercomputer. The extreme flexibility of the cloud idea is a good sign for the future of IT.

Finally, software-defined cloud centers see the light with the faster maturity and stability of the implementation technologies. To attain the originally envisaged goals, researchers are proposing to incorporate software wherever needed in order to bring in the desired separations and sophistications so that significantly higher utilization level can be attained. When the utilization rate goes up, the cost is bound to come down. In short, the longstanding target of infrastructure programmability can be met with the embedding of intelligent software so that the infrastructure manageability, serviceability, and sustainability tasks become easier, economical and quicker.

2.16 Data analytics

With the amount of digital data getting generated every day is projected to be in the range of exabytes, it is important for any corporate to collect, cleanse and crunch the data in order to be digitally transformed. We cannot afford to lose any data anymore. All kinds of internal as well as external data have to be gathered and processed in order to extricate actionable insights out of data heaps. Machine, device, sensor, actuator, business, social, operational, transactional and analytical data have to be consciously gleaned and subjected to a variety of deeper investigations in time to uncover useful and usable insights out of data volumes, Data are predominantly multi-structured. Also, data are categorized as big, fast and streaming data. Batch processing of big data has been the norm thus far. With the technological advancements, big data also can be processed and mined quickly. That is, real-time analytics of big data is made possible. Further on, we have fast and streaming data and they are analyzed in time as the timeliness and trustworthiness of data are essential to take correct decisions. Data typically lose its value with time. Thus, real-time analytics becomes dominant and deft.

As we all know, the big data paradigm is opening up a fresh set of opportunities for businesses. The key challenge in front of businesses is how efficiently and rapidly to capture, process, analyze and extract tactical, operational as well as strategic insights in time to act upon swiftly with all the confidence and clarity. In the recent past, there are experiments using the emerging concept of in-memory computing. For a faster generation of insights out of a massive amount of multi-structured data, the new entrants such as in-memory and in-database analytics are highly reviewed and recommended. The new mechanism insists on putting all incoming data in-memory instead of storing it in local or remote databases so that the major barrier of data latency gets eliminated.

Big Data Analytics (BDA)—The big data paradigm has become a big topic across nearly every business domain. IDC defines big data computing as a set of new-generation technologies and architectures, designed to economically extract value from very large volumes of a wide variety of data by enabling high-velocity capture, discovery, and/or analysis. There are three core components in big data: the data itself, the analytics of the data captured and consolidated, and the articulation of insights oozing out of data analytics processes. There are robust products and services that can be wrapped around one or all of these big data elements. Thus, there is a direct connectivity and correlation between the digital universe and the big data idea sweeping the entire business scene. The vast majority of new data being generated as a result of digitization is unstructured or semi-structured. This means there is a need arising to somehow characterize or tag such kinds of multi-structured big data to be useful and usable. This empowerment through additional characterization or tagging results in metadata, which is one of the fastest-growing sub-segments of the digital universe through metadata itself is a minuscule part of the digital universe. IDC believes that by 2020, a third of the data in the digital universe (more than 13,000 exabytes) will have big data value, only if it is tagged and analyzed. There will be routine, repetitive, redundant data and hence not all data is necessarily useful for big data analytics. However, there are some specific data types that are princely ripe for big analysis such as:

Surveillance Footage—Generic metadata (date, time, location, etc.) is automatically attached to video files. However as IP cameras continue to proliferate, there is a greater opportunity to embed more intelligence into the camera on the edges so that footage can be captured, analyzed, and tagged in real time. This type of tagging can expedite crime investigations for security insights, enhance retail analytics for consumer traffic patterns and of course improve military intelligence as videos from drones across multiple geographies are compared for pattern correlations, crowd emergence, and response or measuring the effectiveness of counterinsurgency.

Embedded and Medical Devices—In future, sensors of all types including those that may be implanted into the body will capture vital and non-vital biometrics, track medicine effectiveness, correlate bodily activity with health, monitor potential outbreaks of viruses, etc., all in real time thereby realizing automated healthcare with prediction and precaution.

Entertainment and Social Media—Trends based on crowds or massive groups of individuals can be a great source of big data to help bring to market the "next big thing," help pick winners and losers in the stock market, and even predict the outcome of elections all based on information users freely publish through social outlets.

Consumer Images—We say a lot about ourselves when we post pictures of ourselves or our families or friends. A picture used to be worth a thousand words but the advent of big data has introduced a significant multiplier. The key will be the introduction of sophisticated tagging algorithms that can analyze images either in real time when pictures are taken or uploaded or en masse after they are aggregated from various websites.

Data empowers Consumers—Besides organizations, digital data helps individuals to navigate the maze of modern life. As life becomes increasingly complex and intertwined, digital data will simplify the tasks of decision-making and actuation. The growing uncertainty in the world economy over the last few years has shifted many risk management responsibilities from institutions to individuals. In addition to this increase in personal responsibility, other pertinent issues such as life insurance, health care, retirement, etc., are growing evermore intricate increasing the number of difficult decisions we all make very frequently. The data-driven insights come handy in difficult situations for consumers to wriggle out. Digital data, hence, is the foundation and fountain of the knowledge society.

Power shifts to the Data-driven Consumers—Data is an asset for all. Organizations are sagacious and successful in promptly bringing out the premium and people-centric offerings by extracting operational and strategically sound intelligence out of accumulated business, market, social, and people data. There is a gamut of advancements in data analytics in the form of unified platforms and optimized algorithms for efficient data analysis, etc. There are plenty of data virtualization and visualization technologies. These give customers enough confidence and ever-greater access to pricing information, service records and specifics on business behavior and performance. With the new-generation data analytics being performed easily and economically in cloud platforms and transmitted to smartphones, the success of any enterprise or endeavor solely rests with knowledge-empowered consumers.

Consumers delegate tasks to Digital Concierges—We have been using a myriad of digital assistants (tablets, smartphones, wearables, etc.) for a variety of purposes in our daily life. These electronics are of great help and crafting applications and services for these specific as well as generic devices empower them to be more right and relevant for us. Data-driven smart applications will enable these new-generation digital concierges to be expertly tuned to help us in many things in our daily life.

As articulated above, there are integrated platforms and databases for performing real-time analytics on big data. Timeliness is an important factor for information to be beneficially leveraged. The appliances and hyper converged infrastructures (HCIs) are in general high-performing, thus guaranteeing higher throughput in all they do. Here too, considering the need for real-time emission of insights, several product vendors have taken the route of software as well as hardware appliances for substantially accelerating the speed with which the next-generation big data analytics get accomplished. A sample of how cloud-based data analytics contributes for establishing and sustaining smarter homes is depicted through the diagram below.

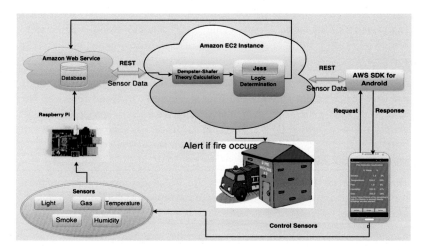

In the business intelligence (BI) domain, apart from realizing real-time insights, analytical processes and platforms are being tuned to bring forth insights that invariably predict something to happen for businesses in the near future. All these advancements enable executives and other stakeholders proactively and pre-emptively to formulate well-defined schemes and action plans, fresh policies, new product offerings, premium services, viable and value-added solutions based on the data-driven insights. Prescriptive analytics, on the other hand, is to assist business executives in prescribing and formulating ways and means of achieving what is predicted.

IBM has introduced a new computing paradigm "stream computing" in order to capture streaming and event data on the fly and to come out with usable and reusable patterns, hidden associations, tips, alerts and notifications, impending opportunities as well as threats, etc., in time for executives and decision-makers to contemplate appropriate countermeasures [1]. The table below clearly tells why cloud centers are efficient and effective for doing IoT data analytics.

- *Agility and Affordability*—No capital investment of large-size infrastructures for analytical workloads. Just use and pay. Quickly provisioned and decommissioned once the need goes down.

- *Data Analytics Platforms in Clouds*—Therefore leveraging cloud-enabled and ready platforms (generic or specific, open or commercial-grade, etc.) are fast and easy

- *NoSQL and NewSQL Databases and Data Warehouses in Clouds*—All kinds of database management systems and data warehouses in cloud speed up the process of next-generation data analytics. Database as a service (DaaS), data warehouse as a service (DWaaS), business process as a service (BPaaS) and other advancements lead to the rapid realization of analytics as a service (AaaS).

- *WAN Optimization Technologies*—There are WAN optimization products for quickly transmitting large quantities of data over the Internet infrastructure

- *Social and professional networking sites* are running in public cloud environments

- *Enterprise-class Applications in Clouds*—All kinds of customer-facing applications are cloud-enabled and deployed in highly optimized and organized cloud environments

- *Anytime, anywhere, any network and any device information and service access* is being activated through cloud-based deployment and delivery

- *Cloud Integrators, Brokers and Orchestrators*—There are products and platforms for seamless interoperability among geographically distributed cloud environments. There are collaborative efforts toward federated clouds and the Intercloud.

- *Sensor/Device-to-Cloud Integration Frameworks* are available to transmit ground-level data to cloud storages and processing.

2.17 Artificial intelligence (AI)

There are other noteworthy developments in the enigmatic IT space. We discussed about big data comprising both historical and current data. All kinds of deterministic and diagnostic analytics are being realized through big and real-time data analytics platforms. There are techniques and tips galore for extracting useful information out of big data in time.

With machine and deep learning (ML/DL) algorithms (a part of the AI discipline), we are heading toward the era of prognostic, predictive, prescriptive and personalized insights out of big data. Clustered computers in conjunction with pioneering ML and DL algorithms can pierce through data heaps to bring forth something useful for people. Primarily, prediction and prescription become a new normal. The domain of computer vision gets a strong boost with the general availability of DL algorithms. Similarly, the natural language processing (NLP) discipline is gaining a lot of attention these days due to the path-breaking ML and DL algorithms. With the explosion of IoT data, the various improvisations in the AI space are to come handy in unambiguously understanding, continuously learning, expertly reasoning out and proposing new thesis out of IoT data. Thus, building and deploying cognitive systems and services become easier with the convergence of IoT and AI concepts. AI algorithms are capable of doing real-time analytics on all kinds of data emanating from different and distributed sources. The growing maturity of the AI domain and the faster proliferation of connected devices in our everyday places have laid down a sustainable foundation toward smarter systems, networks and environments.

2.18 Edge/Fog computing

As accentuated before, there are a plenty of resource-constrained devices in our environments. Similarly, we are being bombarded with resource-intensive devices in our personal, social and professional locations. The brewing trend is that clustering different resource-intensive devices to form a kind of cloud for acquiring multi-structured data from resource-constrained devices in that environment. Thereafter, processing the collected and cleansed data in order to take quick and correct decisions is becoming the new normal. Here comes the various and valid reasons why data analytics of device data has to be accomplished through edge/fog device clouds.

- *Volume and Velocity*—ingesting, processing and storing such huge amounts of data which is gathered in real-time.
- *Security*—devices can be located in sensitive environments, control vital systems or send private data. With the number of devices and the fact they are not humans who can simply type a password, new paradigms and strict authentication and access control must be implemented.
- *Bandwidth*—if devices constantly send the sensor and video data, it will hog the internet and cost a fortune. Therefore, edge analytics approaches must be deployed to achieve scale and lower response time.
- Real-time Data Capture, Storage, Processing, Analytics, Knowledge Discovery, Decision-making and Actuation.
- Less Latency and Faster Response.
- Context-Awareness capability.
- *Combining real-time data with historical state*—there are analytics solutions which handle batch quite well and some tools that can process streams without historical context. It is quite challenging to analyze streams and combine them with historical data in real-time.
- *Power consumption*—Cloud computing is energy-hungry and that it is a concern for a low-carbon economy.
- *Data obesity*—In a traditional cloud approach, a huge amount of untreated data is pumped blindly into the cloud that it is supposed to have magical algorithms written by data scientists. This vision is really not the best efficient and it is much wiser to pre-treat data at a local level and to limit the cloud processes at the strict minimum.

Thus, edge computing and analytics through edge device clouds are flourishing with a burgeoning of industrial and people-centric use cases.

3. Enlisting the recent happenings in the IT space

- *Consumerization—Extended Device Ecosystem*—Trendy and handy, slim and sleek mobile, wearable, implantable and portable devices (Instrumented, Interconnected and Intelligent Devices)

- *Sentient and Smart Materials*—Attaching scores of Digitization/edge technologies (invisible, calm, infinitesimal and disposable sensors and actuators, LEDs, stickers, tags, labels, motes, dots and dust, Beacons, specks, codes, chips, controllers, etc.)
- *Extreme and Deeper Connectivity and Networking Standards*—5G cellular communication for the IoT era, Wi-Fi 802.11 ax, etc.
- *Commoditization and Industrialization—Infrastructure Optimization and Elasticity*—Programmable, Consolidated, Converged, Adaptive, Automated, Shared, QoS-enabling, Green and Lean IT Infrastructures
- Real-time Edge Analytics through Edge/Fog device clouds
- *Compartmentalization through Virtualization and Containerization* through Hypervisors for virtualized workloads and the Docker platform for productive and portable Workloads
- *Middleware Solutions* (Intermediation, Aggregation, Dissemination, Arbitration, Enrichment, Collaboration, Delivery, Management, Governance, Brokering, Identity and Security)
- *In-Memory and In-Database Data Processing Appliances* and integrated platforms for big, fast, and streaming data analytics
- *Process Innovation and Architecture Assimilation* (SOA, EDA, SCA, MDA, ROA, WOA, MSA, etc.)
- *New Kind of Data Sources, Databases and Data Warehouses* (SQL (Clustered, Analytical, and Parallel), NoSQL, NewSQL and Hybrid Models)
- *A Bevy of Pioneering Technologies* (Virtualization, miniaturization, integration, composition, sensing, vision, perception, mobility, knowledge engineering, fog/edge computing, visualization, etc.)
- Natural, Intuitive and Informative Interfaces/Simple (web 1.0), social (web 2.0), semantic (web 3.0) and smart (web 4.0)
- *Artificial, Ambient and Augmented Intelligence* for Predictive, Prescriptive, Personalized and Cognitive Analytics
- Blockchain Technology for Digital Security
- Augmented and Virtual Reality (AR and VR)
- *Programming Languages* (Golang, Scala, RUST, Ballerina, etc.) and Low-code Platforms

There are continuous improvisations and innovations in the technology space. Newer technologies open up fresh possibilities and opportunities for businesses as well as people to embark on deeper and decisive automation. The below diagram pictorially illustrates how devices talking to one another and also with remotely held cloud applications can lead to the realization of numerous sophisticated services.

The Interconnectivity of Devices

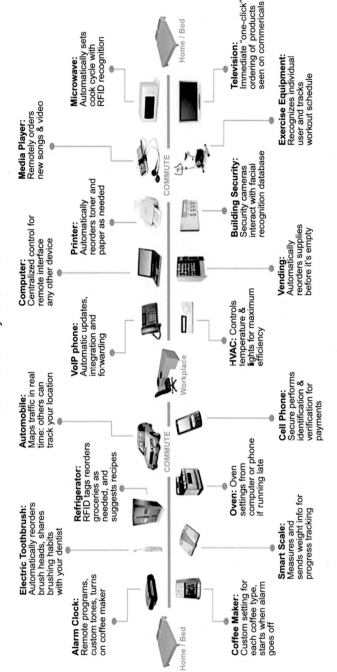

Alarm Clock: Remote programs, custom tones, turns on coffee maker

Electric Toothbrush: Automatically reorders brush heads, shares brushing habits with your dentist

Refrigerator: RFID tags reorders groceries as needed, and suggests recipes

Automobile: Maps traffic in real time; others can track your location

VoIP phone: Automatic updates, integration and forwarding

Computer: Centralized control for remote interface any other device

Printer: Automatically reorders toner and paper as needed

Media Player: Remotely orders new songs & video

Microwave: Automatically sets cook cycle with RFID recognition

Television: Immediate "one-click" ordering of products seen on commericals

Exercise Equipment: Recognizes individual user and tracks workout schedule

Building Security: Security cameras interact with facial recognition database

Vending: Automatically reorders supplies before it's empty

HVAC: Controls temperature & lights for maximum efficiency

Cell Phone: Secure performs identification & verification for payments

Oven: Oven settings from computer or phone if running late

Smart Scale: Measures and sends weight info for progress tracking

Coffee Maker: Custom setting for each coffee type, starts when alarm goes off

Home / Bed

COMMUTE

Workplace

COMMUTE

Home / Bed

4. The connectivity and integration options

It is forecast that there will be a dazzling array of connected devices in large number and zillions of digitized objects in the years ahead. They ought to be seamlessly and spontaneously connected and integrated in order to voluntarily share their unique capabilities with others in order to fulfill complex processes. Here is a gist of integration scenarios.

- Multi-Sensor Fusion—Heterogeneous, multi-faceted, and distributed sensors talk to one another to create sensor mesh to solve complicated problems.
- Sensor to Cloud (S2C) Integration—Cyber-Physical Systems (CPS) will emerge at the intersection of the physical and virtual/cyber worlds.
- Device to Device (D2D) Integration—With the device ecosystem is on the rise, the D2D integration is important.
- Device to Enterprise (D2E) Integration—In order to have remote and real-time monitoring, management, repair, and maintenance, and for enabling decision-support and expert systems, ground-level heterogeneous devices have to be synchronized with control-level enterprise packages such as ERP, SCM, CRM, KM, etc.
- Device to Cloud (D2C) Integration—As most of the enterprise systems are moving to clouds, device to cloud (D2C) connectivity is gaining importance.
- Cloud to Cloud (C2C) Integration—Disparate, distributed and decentralized clouds are getting connected to provide better prospects.
- Mobile Edge Computing (MEC), Cloudlets and Edge Cloud Formation through the clustering of heterogeneous edge/fog devices.

4.1 The big picture

With the cloud space growing fast as the next-generation environment for application design, development, deployment, integration, management, and delivery, the integration scenario is being visualized as pictorially illustrated in the diagram below.

The Big Picture

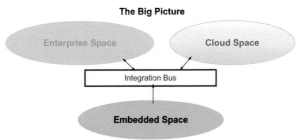

5. The promising digital intelligence methods

We are heading toward the era of zettabytes of digital data. Now it becomes mandatory to apply potential intelligence methods to make sense out of the exponentially growing data. There are new databases, file systems, platforms, algorithms, analytics methods, dashboards and other visualization solutions, message queues, event sources and streams, etc., for quickly and easily squeezing out actionable insights. The list below enumerates the key contributors.

1. Analytics Methods of digital data (Big, Fast, and Streaming data) for diagnostic and deterministic insights.
2. Machine and Deep Learning Algorithms toward predictive, prescriptive and personalized insights.
3. There are powerful data storage solutions such as SQL, NoSQL and NewSQL databases, data warehouses and lakes for digital data analytics.
4. There are in-memory databases for real-time analytics and actuation.
5. There are data virtualization and knowledge visualization tools, platforms, dashboards, etc.
6. There are data processing and analytics platforms (Spark, Storm, Samza, Flink, etc.) on Cloud environments.
7. There are event stores such as Kafka for processing and stocking millions of event messages per second.
8. There are machine and deep learning platforms, frameworks and libraries for accelerating and automating analytics (cognitive analytics).
9. Fog or Edge data analytics is gaining speed with a number of lightweight platform solutions capable of running on fog/edge device clouds.

Here is a macro-level diagram depicting how applications across industry verticals systematically make use of data emanating from multiple sources to be smarter in their actions and reactions. Data analytics platforms and machine learning toolkits on cloud infrastructures contribute enormously for capturing, cleansing, and crunching data to discover and disseminate knowledge. Applications and actuating systems, on getting appropriate insights, exhibit a kind of adaptive and adroit behavior in their delivery. With the process for moving from data to information and to knowledge is maturing fast along with the implementation technologies, data-driven insights and insights-driven decisions will become the new normal for every institution, individual and innovator.

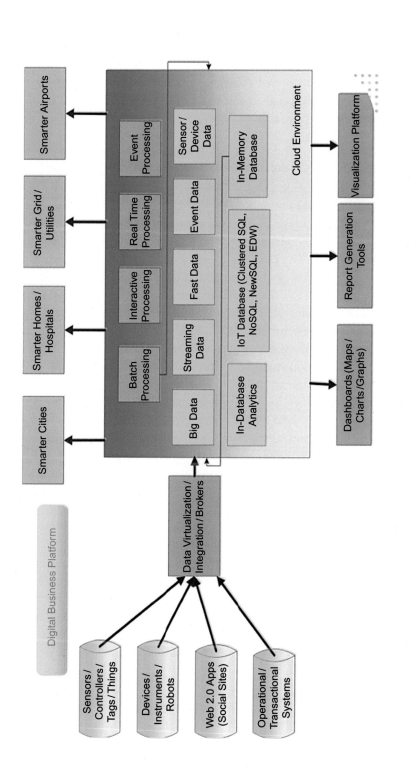

Thus, there are three game changers. Collecting every bit of data, deriving actionable insights out of data volumes, and using the knowledge discovered in a timely and appropriate manner are being touted as the prime factor toward the projected knowledge society.

5.1 The technological approaches toward smarter environments

Our environments are slowly yet steadily stuffed with IoT devices. Any IoT environment typically comprises scores of networked, resource-constrained as well as intensive, and embedded systems. However, IoT artifacts are not individually intelligent. The charter is to make them intelligent individually as well as collectively. Experts have come out with a number of steps to be taken to have intelligent IoT devices. When we have intelligent IoT devices, then the environment altogether is bound to be intelligent. All the occupants and owners of the environment will get a variety of people-centric, situation-aware, and knowledge-filled services. People will be hugely assisted by IoT devices in their everyday tasks. Multiple devices gel well in order to bring forth context-aware, insights-driven, and sophisticated services. As digitized entities and connected devices are being continuously empowered by remotely held software packages, data sources, event streams and knowledge repositories, we can easily anticipate new categories of ground-breaking use cases and applications. In this section, we are to list the prominent and widely accepted ways for empowering our physical, mechanical, electrical and electronics systems to be cognitive.

1. The *Internet of Agents* (*IoA*) for empowering each digital object to be adaptive, articulate, reactive, and cognitive through mapping a software agent for each of the participating digital objects.
2. Through realizing *digital services* (every digital object and connected device is expressed and exposed as a service) and through service orchestration and choreography, business-critical, process-optimized, and situation-aware digital applications can be crafted instantaneously and enhanced accordingly.
3. The emerging concept of *Digital Twin/virtual object representation* is also maturing and stabilizing fast. Many industrial sectors are keenly evaluating and embracing this new paradigm.
4. With the excellent platform support, the proven and potential *IoT data analytics at edge and cloud levels* is the way forward for knowledge discovery and dissemination.

5. The application of *artificial intelligence* (*AI*) technologies (data mining, statistical computing, machine and deep learning algorithms, computer vision, natural language processing, video processing, etc.) leads to the realization of smarter systems, services and solutions.

6. The new concepts of *decentralized applications* and *smart contracts* being popularized through the blockbuster blockchain technology leads to scores of smart and safe applications.

Thus, the connectivity facility being provided by the IoT concept, the cognition capability being the AI algorithms and the security features being realized through the application of the blockchain technology are being pronounced as the key technological paradigms for the future of IT and the society altogether. With intelligent environments abound around us, we will get a number of noteworthy and trustworthy services. We will tend toward digital universe, which brings in a growing array of premium and pioneering competencies for everyone.

5.2 Briefing the brewing idea of digital twin

All kinds of devices in our everyday environments are gradually getting integrated with cyber applications deployed in cloud environments to reap distinct benefits. That is, every physical, mechanical, electrical and electronics systems are being accordingly empowered in order to establish and sustain a seamless and spontaneous integration with faraway cloud applications. This technologically inspired linkage brings in a lot of fresh advantages. For example, manufacturing companies tie their machines at the manufacturing floor with one another as well as with enterprise-scale cloud-hosted applications such as SAP, the leading ERP solution, in order to automate several aspects. We call such enabled systems as cyber-physical systems (CPS). This is becoming common nowadays. The phenomena of the Internet of devices (IoD) and Things (IoT) came around and started to flourish.

Now there is a twist. Increasingly a virtual/logical/digital/cyber version of any physical machine is being created and deposited in cloud environments to be made available online and in an on-demand fashion. The digital version is being blessed with all the features and functionalities of the physical machine. With the cloud idea becoming a core technology, its adoption rate has multiplied in the recent past. There are software-defined and even edge clouds being formed with the latest technologies and tools in order to host and manage cyber applications. Due to their affordability, agility, and availability, the digital versions of physical entities are being made and kept in cloud environments. The physical and digital versions are integrated in order to be constantly in touch to communicate the real-time and runtime data of physical machines.

Besides digital versions, all kinds of historical data, enterprise-class applications, integrated data analytics platforms, machine and deep learning algorithms, etc., are being stocked in distributed cloud environments. Thus comprehensive analytics is easily accomplished. The data flowing from physical machines into their virtual versions enables to do real-time data analytics. Such a linkage enables architects, product engineers and original equipment manufacturers (OEMs) in multiple ways. Before producing a machine physically, all their risks and opportunities can be identified and analyzed fully. How the various system components interact, what are the possible implications, etc., can be proactivity and pre-emptively understood and this knowledge helps designers to come out with competent solutions. Such a scenario is going to be a game-changer for multiple industry verticals.

6. Envisioning the digital universe

The digitization process has gripped the whole world today as never before and its impacts and the associated initiatives are being widely talked about. With an increasing variety of input and output (I/O) devices and newer data sources, the realm of data generation has gone up remarkably. It is forecasted that there will be billions of everyday devices getting connected, capable of generating an enormous amount of data assets which need to be processed. It is clear that the envisaged digital world is to result in a huge amount of bankable data. This growing data richness, diversity, velocity, viscosity, virtuosity, value, and reach decisively activated business organizations and national governments. Thus, there is a fast-spreading of newer terminologies such as digital enterprises and economy. Now it is fascinating the whole world and this new world order has tellingly woken up worldwide professionals and professors to formulate and firm up flexible and futuristic strategies toward digitally transformed business, hotels, retail stores, healthcare, agriculture, manufacturing, etc. There are product vendors, service organizations, research labs, independent software vendors (ISVs), system integrators, consulting companies, etc., are formulating viable technologies, tools, platforms, and infrastructures to tackle this colossal yet cognitive challenge head on. Also, cloud service providers are setting up software-defined compute, networking and storage facilities. Newer types of databases, distributed file systems, data warehouses, data lakes, etc., are being realized to stock up the growing volume of business, personal, machine, people and online data. These data storage solutions ultimately enable specific types of data processing, mining, and analyzing the data getting collected. This pivotal phenomenon has become a clear reason for envisioning the digital universe.

There will be a litany of hitherto unforeseen applications being built and deployed to empower people to experience the digital universe in which all kinds of data producers, middleware, and pre-processing systems, transactional and operational databases, analytical systems, virtualization and visualization tools and software applications will be meaningfully connected with one another. Especially there is a series of renowned and radical transformations in the sensor space. Nanotechnology and other miniaturization technologies have brought in legendary changes in sensor design. The nanosensors can be used to detect vibrations, motion, sound, color, light, humidity, chemical composition and many other characteristics of their deployed environments. These sensors can revolutionize the search for new oil reservoirs, structural integrity for buildings and bridges, merchandise tracking and authentication, food and water safety, energy use and optimization, healthcare monitoring and cost savings, and climate and environmental monitoring. The point to be noted here is the volume of real-time data being emitted by the army of sensors and actuators is exponentially growing and with the help of real-time analytics platforms and algorithms, real-time insights get squeezed out and supplied to actuating devices and people to ponder about and execute right counter measures in time.

The steady growth of sensor networks increases the need for 1 million times more storage and processing power by 2020. It is projected that there will be one trillion sensors by 2025 and every single person will be assisted by approximately 150 sensors in this planet. Cisco has predicted that there will be 50 billion connected devices in 2020 and hence the days of the Internet of Everything (IoE) are not too far off. All these scary statistics convey one thing. That is, IT applications, services, platforms, and infrastructures need to be substantially and smartly invigorated to meet up all sorts of business and peoples' needs in the ensuing era of deepened digitization.

Precisely speaking, the data volume is going to be humongous as the digitization prospect is growing deep and wide. The resulting digitization-induced digital universe and economy will, therefore, be at war with the amount of data being collected and analyzed. The data complexity through the data heterogeneity and multiplicity will be a real challenge and concern for enterprise IT teams. That is, as accentuated above, the real-time analytics of big data is to lead to a series of disruptions and transformations.

7. Conclusion

There have been a number of disruptive and decisive transformations in the information technology (IT) discipline in the last five decades. Silently and succinctly IT has moved from a specialized tool to become a greatly and gracefully leveraged tool on nearly every aspect of our lives today. For a long time, IT has been a business enabler and is steadily on the way to becoming the people enabler in the years to unfold. Once upon a time IT was viewed as a cost center and now the scene has totally changed. That is, IT has become a profit center for all kinds of organizations across the globe. The role and relevance of IT in this deeply connected the world are really phenomenal. IT continuously penetrates into newer domains to bring in automation and acceleration. IT has the innate strength in seamlessly capturing and embodying all kinds of proven and potential innovations and improvisations in order to sustain its marvelous and mesmerizing journey for the betterment of the human society.

The brewing trends clearly vouch for a digital universe in the years ahead. The distinct characteristic of the digitized universe is nonetheless the huge data collection (big data) from a variety of sources. This voluminous data production and the clarion call for squeezing out workable knowledge out of the data for adequately empowering the total human society is activating IT experts, engineers, evangelists, and exponents to incorporate more subtle and successful disruptions and transitions in the IT field. The slogan "more with less" is becoming louder. The inherent expectations from IT for resolving various social, business, and personal problems are on the climb. In this chapter, we have discussed about the digitization paradigm and how next-generation cloud environments are to support and sustain the digital universe. Lastly, we have introduced about the origin and the journey of the concept.

References

[1] Kobielus J: *The Role of Stream Computing in Big Data Architectures,* Retrieved from, http://ibmdatamag.com/2013/01/the-role-of-stream-computing-in-big-data-architectures/, 2013.

Further reading

[2] Devlin B: *The Big Data Zoo—Taming the Beasts, a White Paper by 9sight Consulting,* Retrieved from, http://ibmdatamag.com/2012/10/the-big-data-zoo-taming-the-beasts/, 2012.

[3] *Businesses are Ready for a New Approach to IT, a White Paper From IBM,* Retrieved from,
     http://api.ning.com/files/4O3HllTJ*6eYzuJojtOShe1qdgXURtFikZkbms0xSclfpF
     5tysAZqSZiC9k3N-4iggw2uxSnW5rFxf6MKEMTJxzbSvdYfF26/Businessesareready for
     anewapproachtoIT.pdf, 2012.
[4] *Intuit: The new data democracy, a 2020 report,* Retrieved fromhttp://network.intuit.com/
     wp-content/uploads/2012/12/intuit_corp_vision2020_121412-final.pdf, 2013.
[5] Gartner: *Gartner: Top 10 Strategic Technology Trends for 2014,* Retrieved
     from, http://www.forbes.com/sites/peterhigh/2013/10/14/gartner-top-10-strategic-
     technology-trends-for-2014/, 2013.
[6] Gantz J, Reinsel D: *The Digital Universe in 2020: Big Data, Bigger Digital Shad-Ows, and
     Biggest Growth in the Far East, an IDC View,* Retrieved from, http://www.emc.com/
     leadership/digital-universe/iview/index.htm, 2012.
[7] Vodafone: *Connecting to the Cloud: Business Advantage From Cloud Services, a White Paper,*
     Retrieved   from,   http://www.vodafone.co.nz/a/pdfs/corporate-and-government/
     vf-white-paper-connect-cloud.pdf, 2010.
[8] Acosta A: *Evolving Shared Infrastructure in Large Data Centers, a White Paper,* Retrieved from,
     http://www.dell.com/learn/us/en/04/business~solutions~power~en/documents~ps1q
     13-20130197-bartleson.pdf, 2013.
[9] Intel: *Distributed Data Mining and Big Data, a Vision paper,* retrieved from, http://www.
     intel.com/content/www/us/en/big-data/distributed-data-mining-paper.html, 2013.
[10] HP: *HP Converged Infrastructure, a Brochure,* Retrieved from, http://h17007.www1.hp.
     com/in/en/converged-infrastructure/, 2011.
[11] HP: *Information Optimization, a Business White Paper,* retrieved from, http://www8.hp.
     com/us/en/business-solutions/big-data.html, 2012.

About the authors

Pethuru Raj working as the Chief Architect in the Site Reliability Engineering (SRE) division, Reliance Jio Infocomm Ltd. (RJIL), Bangalore. The previous stints are in IBM Cloud Center of Excellence (CoE), Wipro Consulting Services (WCS), and Robert Bosch Corporate Research (CR). In total, I have gained more than 18 years of IT industry experience and 8 years of research experience. Finished the CSIR-sponsored PhD at Anna University, Chennai and continued with the UGC-sponsored postdoctoral research in the Department of Computer Science and Automation, Indian Institute of Science, Bangalore. Thereafter, I was granted a couple of international research fellowships (JSPS and JST) to work as a Research Scientist for 3.5 years in two leading Japanese universities. Published more than 30 research papers in peer-reviewed journals such as IEEE, ACM, Springer-Verlag, Inderscience, etc. Have authored and edited 20 books thus far and focus on some of the emerging technologies such as IoT, Cognitive Analytics, Blockchain, Digital Twin, Docker-enabled Containerization, Data Science, Microservices Architecture, fog/edge computing, Artificial intelligence (AI), etc. Have contributed 35 book chapters thus far for various technology books edited by highly acclaimed and accomplished professors and professionals.

Jenn-Wei Lin is a full professor in the Department of Computer Science and Information Engineering, Fu Jen Catholic University, Taiwan. He received the Ph.D. degree in electrical engineering from National Taiwan University, Taiwan, in 1999. His research interests are cloud computing, mobile computing and networks, distributed systems, and fault-tolerant computing.

CHAPTER TWO

Digital twin technology for "smart manufacturing"

Preetha Evangeline[a], Anandhakumar[b]
[a]Karunya Institute of Technology and Sciences, Coimbatore, Tamil Nadu, India
[b]Anna University, Chennai, Tamil Nadu, India

Contents

Abstract

The insights into additive manufacturing, the Internet of Things, and analytics enable us to help organizations reassess their people, processes, and technologies in light of advanced manufacturing practices that are evolving every day. Manufacturing processes are becoming increasingly digital. As this trend unfolds, many companies often struggle to determine what they should be doing to drive and deliver real value both operationally and strategically. Indeed digital solutions may promise significant value for an organization. Of particular fascination of late seems to be the notion of a digital twin: a near-real-time digital image of a physical object or process that helps optimize business performance.

Advances in Computers, Volume 117
ISSN 0065-2458
https://doi.org/10.1016/bs.adcom.2019.10.009

1. Introduction

Until recently, the digital twin and the massive amounts of data it processes often remained elusive to enterprises due to limitations in digital technology capabilities as well as prohibitive computing, storage, and bandwidth costs. Such obstacles, however, have diminished dramatically in recent years [1]. Significantly lower costs and improved power and capabilities have led to exponential changes that can enable leaders to combine information technology (IT) and operations technology (OT) to enable the creation and use of a digital twin [2]. So why is the digital twin so important, and why should organizations consider it? The digital twin can allow companies to have a complete digital footprint of their products from design and development through the end of the product life cycle. This, in turn, may enable them to understand not only the product as designed but also the system that built the product and how the product is used in the field. With the creation of the digital twin, companies may realize significant value in the areas of speed to market with a new product, improved operations, reduced defects, and emerging new business models to drive revenue.

The digital twin may enable companies to solve physical issues faster by detecting them sooner, predict outcomes to a much higher degree of accuracy, design and build better products, and, ultimately, better serve their customers. With this type of smart architecture design, companies may realize value and benefits iteratively and faster than ever before. It can be a daunting task to create a digital twin if a company would like to try this all at once. The key could be to start in one area, deliver value there, and continue to develop. But before anything else, enterprises should first understand the definition of and approach to the development of the digital twin in order to avoid being overwhelmed. In the pages that follow, we discuss the digital twin—its definition, the way it can be created, how it could drive value, its typical applications in the real world, and how a company can prepare for the digital twin planning process.

2. Digital twin, what it is?

Industry and academia define a digital twin in several different ways. However, perhaps neither group places the required emphasis on the process aspects of a digital twin. For example, according to some, a digital twin is an integrated model of an as-built product that is intended to reflect all manufacturing defects and be continually updated to include the wear and tear sustained while in use [3]. Other widely circulated definitions describe the digital twin as a sensor-enabled digital model of a physical object that simulates the object in a live setting [4].

A digital twin can be defined, fundamentally, as an evolving digital profile of the historical and current behavior of a physical object or process that helps optimize business performance. The digital twin is based on massive, cumulative, real-time, real-world data measurements across an array of dimensions. These measurements can create an evolving profile of the object or process in the digital world that may provide important insights on system performance, leading to actions in the physical world such as a change in product design or manufacturing process. A digital twin differs from traditional computer aided design (CAD), nor does it serve as merely another sensor-enabled Internet of Things (IoT) solution [5]. It could be much more than either. CAD is completely encapsulated in a computer-simulated environment that has demonstrated moderate success in modeling complex environments; [6] and more simple IoT systems measure things such as position and diagnostics for an entire component, but not interactions between components and the full life cycle processes [7].

Indeed, the real power of a digital twin—and why it could matter so much is that it can provide a near-real-time comprehensive linkage between the physical and digital worlds. It is likely because of this interactivity between the real and digital worlds of product or process that digital twins may promise richer models that yield more realistic and holistic measurements of unpredictability. And thanks to cheaper and more powerful computing capabilities, these interactive measurements can be analyzed with modern-day massive processing architectures and advanced algorithms for real-time predictive feedback and offline analysis. These can enable fundamental design and process changes that would almost certainly be unattainable through current methods. A digital twin can be defined, fundamentally, as an evolving digital profile of the historical and current behavior of a physical object or process that helps optimize business performance.

Digital twins are designed to model complicated assets or processes that interact in many ways with their environments for which it is difficult to predict outcomes over an entire product life cycle [8]. Indeed, digital twins may be created in a wide variety of contexts to serve different objectives. For example, digital twins are sometimes used to simulate specific complex deployed assets such as jet engines and large mining trucks in order to monitor and evaluate wear and tear and specific kinds of stress as the asset is used

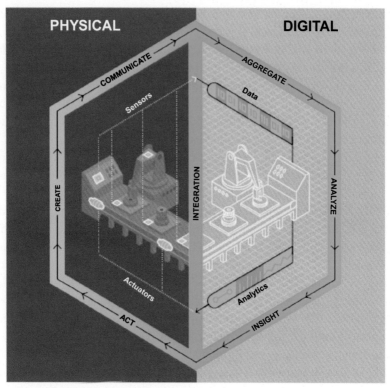

Fig. 1 Manufacturing process digital twin model.

in the field. Such digital twins may yield important insights that could affect future asset design. A digital twin of a wind farm may uncover insights into operational inefficiencies. Other examples of deployed asset-specific digital twins abound [8]. As insightful as digital twins of specific deployed assets may be, the digital twin of the manufacturing process appears to offer an especially powerful and compelling application. Fig. 1 represents a model of a manufacturing process in the physical world and its companion twin in the digital world. The digital twin serves as a virtual replica of what is actually happening on the factory floor in near-real-time. Thousands of sensors distributed throughout the physical manufacturing process collectively capture data along a wide array of dimensions: from behavioral characteristics of the productive machinery and works in progress (thickness, color qualities, hardness, torque, speeds, and so on) to environmental conditions within the factory itself. These data are continuously communicated to and aggregated by the digital twin application. The digital twin application continuously analyses incoming data streams. Over a period of time, the analyses may uncover unacceptable trends in the actual performance of the manufacturing process in a particular dimension when compared with an ideal range of tolerable performance. Such comparative insight could trigger investigation and a potential change to some aspect of the manufacturing process in the physical world. This is the journey of interactivity between the physical and digital worlds, which Fig. 1 endeavors to convey. Such a journey underscores the profound potential of the digital twin: thousands of sensors taking continuous, nontrivial measurements that are streamed to a digital platform, which, in turn, performs near-real-time analysis to optimize a business process in a transparent manner. The model of Fig. 1 specifically finds expression through five enabling components—sensors and actuators from the physical world, integration, data, and analytics—as well as the continuously updated digital twin application. These constituent elements of Fig. 1 are explained at a high level below:

2.1 Sensors

Sensors distributed throughout the manufacturing process create signals that enable the twin to capture operational and environmental data pertaining to the physical process in the real world.

2.2 Data

Real-world operational and environmental data from the sensors are aggregated and combined with data from the enterprise, such as the bill of materials (BOM), [9] enterprise systems, and design specifications. Data may also contain other items such as engineering drawings, connections to external data feeds, and customer complaint logs.

3. Integration

Sensors communicate the data to the digital world through integration technology (which includes edge, communication interfaces, and security) between the physical world and the digital world, and vice versa.

3.1 Analytics

Analytics techniques are used to analyze the data through algorithmic simulations and visualization routines that are used by the digital twin to produce insights.

3.2 Digital twin

The "digital" side of Fig. 1 is the digital twin itself—an application that combines the components above into a near-real-time digital model of the physical world and process. The objective of a digital twin is to identify intolerable deviations from optimal conditions along any of the various dimensions. Such a deviation is a case for business optimization; either the twin has an error in the logic (hopefully not), or an opportunity for saving costs, improving quality, or achieving greater efficiencies has been identified. The resulting opportunity may result in an action back in the physical world.

3.3 Actuators

Should an action be warranted in the real world, the digital twin produces the action by way of actuators, subject to human intervention, which trigger the physical process [10]. Clearly, the world of a physical process (or object) and its digital twin analogue are vastly more complex than a single model or framework can depict. And, of course, the model of Fig. 1 is just one digital twin configuration that focuses on the manufacturing portion of the product life cycle [11]. But what our model aims to show is the integrated, holistic, and iterative quality of the physical and digital world pairing. It is through that prism that one may begin the actual process that serves to create a digital twin.

Digital twin process design and information requirements.

In general, the creation of the digital twin encompasses two main areas of concern:

1. Designing the digital twin processes and information requirements in the product life cycle from the design of the asset to the field use and maintenance of the asset in the real world.
2. The creation of the enabling technology to integrate the physical asset and its digital twin for real-time flow of sensor data and operational and transactional information from the company's core systems, as expressed in a conceptual architecture.

The digital twin creation starts with process design. What are the processes and integration points for which the twin will be modeling? Standard process design techniques should be used to show how business processes, people enabling the processes, business applications, information, and physical assets interact. Diagrams are created that link the process flow to the applications, data needs, and the types of sensor information required to create the

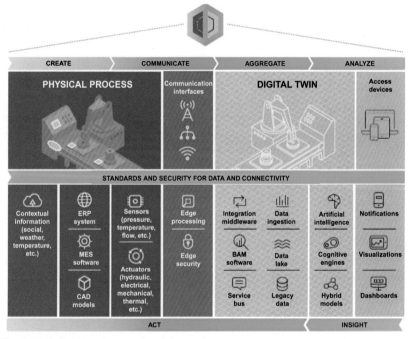

Fig. 2 Digital twin conceptual architecture.

digital twin. The process design is augmented with attributes where cost, time, or asset efficiency could be improved. These typically form the base line assumptions from which the digital twin enhancements should begin.

The digital twin conceptual architecture (Fig. 2) can rightly be thought of as an expansive or "under the hood" look at the enabling components that comprise the manufacturing process digital twin model of Fig. 1, although the same basic principles may likely apply in any digital twin configuration. The conceptual architecture may be best understood as a sequence of six steps, as follows:

1 *Create*: The create step encompasses outfitting the physical process with myriad sensors that measure critical inputs from the physical process and its surroundings. The measurements by the sensors can be broadly classified into two categories: (1) operational measurements pertaining to the physical performance criteria of the productive asset (including multiple works in progress), such as tensile strength, displacement, torque, and color uniformity; (2) environmental or external data affecting the operations of a physical asset, such as ambient temperature, barometric pressure, and moisture level. The measurements can be transformed into secured digital messages using encoders and then transmitted to the digital twin. The signals from the sensors may be augmented with process-based information from systems such as the manufacturing execution systems, enterprise resource planning systems, CAD models, and supply chains systems. This would provide the digital twin with a wide range of continually updating data to be used as input for its analysis.

2 *Communicate*: The communicate step helps the seamless, real-time, bidirectional integration/connectivity between the physical process and the digital platform. Network communication is one of the radical changes that have enabled the digital twin; it comprises three primary components:

  a. Edge processing: The edge interface connects sensors and process historians, processes signals and data from them near the source, and passes data along to the platform. This serves to translate proprietary protocols to more easily understood data formats as well as reduce network communication. Major advances in this area have eliminated many bottlenecks that have limited the viability of a digital twin in the past.

  b. Communication interfaces: Communication interfaces help transfer information from the sensor function to the integration function. Many options are needed in this area, given that the sensor producing the insight can, in theory, be placed at almost any location, depending on the digital twin configuration under consideration: inside a factory, in a home, in a mining operation, or in a parking lot, among myriad other locations.

  c. Edge security: New sensor and communication capabilities have created new security issues, which are still developing. The most common security approaches are to use firewalls, application keys, encryption, and device certificates. The need for new solutions to safely enable digital twins will likely become more pressing as more and more assets become IP enabled.

3 *Aggregate*: The aggregate step can support data ingestion into a data repository, processed and prepared for analytics. The data aggregation and processing may be done either on the premises or in the cloud. The technology domains that power data aggregation and processing have evolved tremendously over the last few years in ways that allow designers to create massively scalable architectures with greater agility and at a fraction of the cost in the past.

4 *Analyze*: In the analyze step, data is analyzed and visualized. Data scientists and analysts can utilize advanced analytics platforms and technologies to develop iterative models that generate insights and recommendations and guide decision making.

5 *Insight*: In the insight step, insights from the analytics are presented through dashboards with visualizations, highlighting unacceptable differences in the performance of the digital twin model and the physical world analogue in one or more dimensions, indicating areas that potentially need investigation and change.

6 *Act*: The act step is where actionable insights from the previous steps can be fed back to the physical asset and digital process to achieve the impact of the digital twin. Insights pass through decoders and are then fed into the actuators on the asset process, which are responsible for movement or control mechanisms, or are updated in back-end systems that control supply chains and ordering behavior all subject to human intervention. This interaction completes the closed loop connection between the physical world and the digital twin. The digital twin application is usually written in the primary system language of the enterprise, which uses the above steps to model the physical asset and processes. In addition, throughout the process, standards and security measures may be applied for purposes of data management and interoperable connectivity. The computation power of big data engines, the versatility of the analytics technologies, the massive and flexible storage possibilities of the aggregation area, and integration with canonical data allow the digital twin to model a much richer, less isolated environment than ever before. In turn, such developments may lead to a more sophisticated and realistic model, all with the potential of lower-cost software and hardware. It is important to note that the above conceptual architecture should be designed for flexibility and scalability in terms of analytics, processing, the number of sensors and messages, etc. This can allow the architecture to evolve rapidly with the continual, and sometimes exponential, changes in the market.

4. Digital twin business values

With the emergence of increasingly favorable storage and computing costs, the number of use cases and possibilities to enable a digital twin has greatly expanded, in turn driving business value. When considering the business value that the digital twin offers, companies should focus on issues related to strategic performance and marketplace dynamics, including improved and longer-lasting product performance, faster design cycles, potential for new revenue streams, and better warranty cost management. These strategic issues, among others, can translate into specific applications that might afford broad business value that a digital twin may realize. Table 1 lists a summary of such values by category. Besides the areas of business values mentioned above, a digital twin may help address many other key performance and efficiency metrics for a manufacturing company. Overall, the digital twin may offer many applications to drive value and start to fundamentally change how a company does business. Such value may be measured in tangible results that may be tracked back to key metrics for a business.

Table 1 Digital twin business values.

Category of business value	Potential specific business values
Quality	• Improve overall quality • Predict and detect quality trend defects sooner • Control quality escapes and be able to determine when quality issue started
Warranty cost and services	• Understand current configuration of equipment in the field to be able to service more efficiently • Proactively and more accurately determine warranty and claims issues to reduce overall warranty cost and improve customer experiences
Operations cost	• Improve product design and engineering change execution • Improve performance of manufacturing equipment • Reduce operations and process variability
Record retention and serialization	• Create a digital record of serialized parts and raw materials to better manage recalls and warranty claims and meet mandated tracking requirements
New product introduction cost and lead time	• Reduce the time to market for a new product • Reduce overall cost to produce new product • Better recognize long-lead-time components and impact to supply chain
Revenue growth opportunities	• Identify products in the field that are ready for upgrade • Improve efficiency and cost to service product

Given the wide applications of the digital twin, how does one get started? A major challenge in undertaking a digital twin process can reside in determining the optimal level of detail in creating a digital twin model. While an overly simplistic model may not yield the value a digital twin promises, taking too fast and broad an approach can almost guarantee getting lost in the complexity of millions of sensors, hundreds of millions of signals the sensors produce, and the massive amount of technology to make sense of the model. Therefore, an approach that is either too simplistic or too complex could kill the momentum to move forward. Fig. 3 offers a possible approach that falls somewhere in between.

4.1 Imagine the possibilities

The first step would be to imagine and shortlist a set of scenarios that could benefit from having a digital twin. The right scenario may be different for every organization and circumstance, but will likely have the following two key characteristics:

1. The product or manufacturing process being considered is valuable enough for the enterprise to invest in building a digital twin.
2. There are outstanding, unexplained processor product-related issues that could potentially unlock value either for the customers or the enterprise.

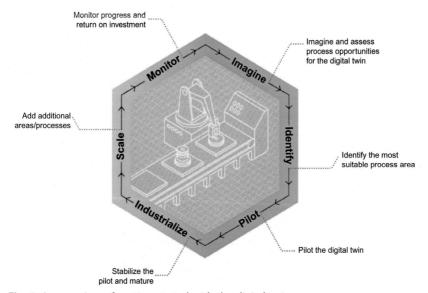

Fig. 3 An overview of getting started with the digital twin.

After the shortlist of scenarios is created, each scenario would be assessed to identify pieces of the process that can provide quick wins by using a digital twin. We encourage a focused ideation session with members of operational, business, and technical leadership for expediting the assessment.

4.2 Identify the process

The next step would be to identify the pilot digital twin configuration that is both of the highest possible value and has the best chance of being successful. Consider operational, business, and organizational change management factors in identifying which configurations could be best candidates for the pilot. Focus on areas that have potential to scale across equipment, sites, or technologies. Companies may face challenges going too deep into a specific digital twin of a highly complex equipment or manufacturing process, while the ability to deploy broadly across the organization tends to drive the most value and support: Focus on going broad rather than deep.

4.3 Pilot a program

Consider moving quickly into a pilot program using iterative and agile cycles to accelerate learning, manage risk proactively, and maximize return on initial investments. The pilot can be a subset of business divisions, or products to limit scope, but with the ability to show value to the enterprise. As you move through the pilot, the implementation team should support adaptability and an open mind-set at any time of your journey, maintain an open and agnostic ecosystem that would allow adaptability and integration with new data (structured and unstructured) and leverage new technologies or partners. While you should want to be agnostic to any type of data sources (for example, new sensors and external data sources), you also need a solution that can support the expansion of an end-to-end solution (from early development to after sales). As soon as the initial value is delivered, consider building on this momentum to continue the drive for greater results. Communicate the value realized to the larger enterprise.

4.4 Industrialize the process

Once success is shown in the field, you can industrialize the digital twin development and deployment process using established tools, techniques, and playbooks. Manage expectations from the pilot team and other projects seeking to adopt it. Develop insights on the digital twin process and publish to the larger enterprise. This may include moving from a more siloed implementation to integration into the enterprise, implementation of a data lake, performance and throughput enhancements, improved governance and data standards, and implementation of organizational changes to support the digital twin.

4.5 Scale the twin

Once successful, it can be important to identify opportunities to scale the digital twin. Target adjacent processes and processes that have interconnections with the pilot. Use the lessons learned from the pilot and the tools, techniques, and playbooks developed during the pilot to scale expeditiously. As you scale, continue to communicate the value realized through the adoption of the digital twin by the larger enterprise and shareholders.

4.6 Monitor and measure

Solutions should be monitored to objectively measure the value delivered through the digital twin. Identify whether there were tangible benefits in cycle time, yield throughput, quality, utilization, incidents, and cost per item, among others. Make changes to digital twin processes iteratively, and observe results to identify the best possible configuration. Most importantly, this is not a project that should typically end once a benefit is identified, implemented, and measured. To continually differentiate in the market place, companies should plan time to move through the cycle again in new areas of the business over time. All in all, true success in achieving early milestones on a digital twin journey will likely rely on an ability to grow and sustain the digital twin initiative in a fashion that can demonstrate increasing value for the enterprise over time. To help ensure such an outcome, one may need to integrate digital technologies and the digital twin into the complete organizational structure from R&D to sales continuously leveraging digital twin insights to change how the company conducts business, makes decisions, and creates new revenue streams.

5. Conclusion

The digital twin may drive tangible value for companies, create new revenue streams, and help them answer key strategic questions. With new technology capabilities, flexibility, agility and lower cost, companies may be able to start their journeys to create a digital twin with lower capital investment and shorter time to value than ever before. A digital twin has many applications across the life cycle of a product and may answer questions in real time that couldn't be answered before, providing kinds of value considered nearly inconceivable just a few years ago. Perhaps the question is not whether one should get started, but where one should start to get the biggest value in the shortest amount of time, and how one can stay ahead of the competition. What will be the first step, and how will you get started? It can be an overwhelming task to get there, but the journey starts with a single step.

References

[1] Economist: *The Cheap, Convenient Cloud, 2015:* 2015, Economist http://www.economist.com/news/business/21648685-cloud-computing-prices-keep-falling-whole-it-business-will-change-cheap-convenient.

[2] Mussomeli A, Gish D, Laaper S: *The Rise of the Digital Supply Network, 2016:* 2016, Deloitte University Press https://dupress.deloitte.com/dup-us-en/focus/industry-4-0/digital-transformation-in-supplychain.html.

[3] Reid J, Rhodes D: Digital system models: an investigation of the non-technical challenges and research needs. In *Conference on Systems Engineering Research, Systems Engineering Advancement Research Initiative, Massachusetts Institute of Technology*; 2016.

[4] Grieves M: *Digital Twin: Manufacturing Excellence Through Virtual Factory Replication, 2014:* 1. http://innovate.fit.edu/plm/documents/doc_mgr/912/1411.0_Digital_Twin_White_Paper_Dr_Grieves.pdf.

[5] Holdowsky J, et al: *Inside the Internet of Things (IoT): A Primer on the Technologies Building the IoT, 2015:* 2015, Deloitte University Press, https://dupress.deloitte.com/dup-us-en/focus/internet-of-things/iot-primer-iot-technologies-applications.html, For an overview of the IoT and the technologies that are a part of it, see.

[6] West T, Pyster A: Untangling the digital thread: the challenge and promise of model-based engineering in defense acquisition, *Insight* 18(2):45–55, 2015. https://doi.org/10.1002/inst.12022.

[7] Holdowsky et al., 2015, Inside the Internet of Things (IoT) Deloitte University Press.

[8] Grieves, 2016, Digital Twin.

[9] Daecher A, Schmid R: *Internet of Things: From Sensing to Doing, 2016:* Tech Trends, 2016, Deloitte University Press https://dupress.deloitte.com/dup-us-en/focus/tech-trends/2016/internetof-things-iot-applications-sensing-to-doing.html.

[10] Holdowsky et al., Inside the Internet of Things (IoT) A primer on the Technologies building IoT, 2015, Deloitte University Press.

[11] Cotteleer M, Trouton S, Dobner E: *3D Opportunity and the Digital Thread: Additive Manufacturing Ties it all Together, 2016:* 2016, Deloitte University Press https://dupress.deloitte.com/dup-us-en/focus/3dopportunity/3d-printing-digital-thread-in-manufacturing.html.

About the authors

Dr. Preetha Evangeline is currently working as an assistant professor in the Department of Computer Science and Engineering at Karunya Institute of Technology and Sciences, Coimbatore. She holds a PhD from Anna University, Chennai and her doctoral research was in the area of Cloud Computing. Her area of expertise lies in the field of High Performance Computing, Intelligent Computing, Data Structures, and Operating Systems. She works on the recent research field of Process Digital Twin Technology and her upcoming research works will be multidisciplinary collaborating various departments to solve socially related challenges and provide solutions to human problems.

Dr. Anandhakumar is a professor in the Department of Information Technology at Anna University, Chennai. He has completed his doctorate in the year 2006 from Anna University. He has produced 17 PhD's in the field of Image Processing, Cloud Computing, Multimedia technology, and Machine Learning. His ongoing research lies in the field of Digital Twin Technology. He has published more than 150 papers in the reputed journals.

CHAPTER THREE

Using fog computing/edge computing to leverage Digital Twin

J. Pushpa[a] and S.A. Kalyani[b]
[a]Jain University, Bangalore, India
[b]East-West College, Bangalore, India

Contents

Advances in Computers, Volume 117
ISSN 0065-2458
https://doi.org/10.1016/bs.adcom.2019.09.003

Abstract

Recreating a real-world entity as a virtual object has already been studied in most of the field, but making the cloned object more intelligent and healer of real-time physical object will give a new vision on technology. Digital Twin is fitting into the above statement. It uses a combination of machine learning, artificial intelligence, the IoT, and big data to evolve as a ubiquitous solution for all kinds of issues. Digital Twin can build on many form based on the requirement which is basically designed to resolve the challenges of the real world entity.

Digital Twin is not limited to solving issues with standalone systems, single entities and machinery problems; it is also suitable for all kinds of data management and controlling issues. Digital Twin can extend its reach by embedding with edge or fog computing which can reduce connectivity and latency issues in networks.

In this chapter, methodologies for leveraging Digital Twin using fog/edge computing will be discussed along with suitable use cases, such as wind turbines, product management, healthcare centers, and so on. We also discuss the financial benefits, tangible benefits, and connectivity model.

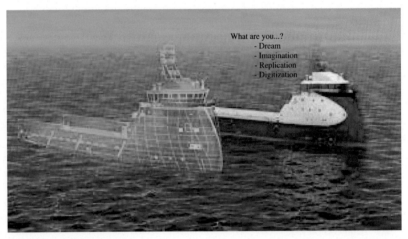

Visualizing a digital twin.

1. Introduction

Digital Twin is a facet of the Internet of Things (IoT) which has emerged in response to new requirements in industries and as a step toward a new digital world. Digital Twin can be defined as a virtual device or monitoring device of a physical entity which can be adopted by all factors in the world.

Between 2002 and 2020, many blogs, articles, and research papers studied the concept of digital twin to understand its potentiality and applicability.

Most cloud vendors are looking forward the cloud based service to digital twin which gives the facilities to share resources, reuse existing components, and create data-centric systems.

The world is moving toward digitization where data transmission is high between users and connected devices, and also many coordination device are increasing as end devices increases. Completing all those requests using computing devices requires high bandwidth and channels, and also a modulator to boost the signal if it approaches the cloud. Cloud computing provides different services such as IaaS, SaaS, and NaaS which can be applied to many industrial settings to facilitate the new technology. With the emergence of new trends and technologies in recent years, cloud computing and the Internet of Things are some of the widely used technologies of today. We have seen tremendous growth in the number of IoT devices and dependencies on these devices.

The ever-increasing growing increase in the number of IoT devices in the industry has led to a huge data deluge. These massive amounts of data coming from IoT devices which are at the edge of the networks need to be processed. For this, huge amounts of data need to be transferred to and from the data centers or the cloud over the network, pushing network bandwidth to the limit. Despite improvements in network technology, data centers cannot guarantee acceptable transfer rates and response times which may be critical for applications. Though the cloud provides different services, the cost is also increased in terms of delay, accessibility, and throughput, due to distance, approach ability and its heavy resources. Digital Twin is a technique commonly implemented on IoT devices which are mostly defined in confined region and are low power devices. Hence cloud computing is not an appropriate solution for those devices. Real-time data processing, easy access, high throughput, low latency, and quick analysis are some of the prime parameters in digital twin and also for real-time devices. Edge computing may be the solution for it. The digital twin framework not only contains the components of physical entity but also integrates with current technology to project the possible cases.

Edge computing is a cloud technology which is locally available in the HAN and PAN which is evolved to achieve prime parameters of networks for real-time systems. In other words, all the processing can be performed at a location which is physically closer to the data source itself, thus leveraging bandwidth, speed, maintenance, automation, etc. In order to achieve this, we need a faster, cheaper, and smarter approach than the traditional approach which typically gathers and then sends data through networks to the cloud or other environments for processing. The edge computing platform provides some capabilities to edge devices which reduce the traffic toward the cloud which in turn reduces the numbers of coordinators and modulators in the network. Edge computing consists of the idea of pushing the computational services or functionalities either completely or partially to the edge of the network is Edge Computing. This chapter discusses edge computing for digital twin, and also fog computing which increases the efficiency and response to edge devices. Fog computing and edge computing are used interchangeably in cloud environments; they look similar but operate differently.

Fog computing is also a cloud technology in the local area network (LAN) which acts like an intermediate layer between edge devices and cloud devices. As discussed by the pod group from *IoT for all* [1], fog computing is a mini cloud within a LAN which builds a data center-like cloud to service the multiple connected nodes in an intra-network.

Manifestation.

In an oracle document [2] regarding the attribute of physical model is mentioned such as name, location, sensor, observed and desired attributes. It is also important to consider other attributes of links such as rate of data transfer, data format, and standards; technology integration techniques should also be specified.

This chapter discusses the characteristics of fog and edge computing along with their services on digital twin, and the comparative studies that identify the benefits of each. We also study a few of the use cases of digital twin and some of the fields within which it can be evolved.

2. Illustrating the epoch-making IoT journey

The mesmerizing number of smart sensors and actuators being deployed in specific environments ultimately produces massive volumes of data. Currently, the collected data is faithfully transmitted over the internet or any private network to faraway cloud infrastructures in order to be concertedly and calculatedly crunched to extract exceptional insights. Clouds are widely considered to be the best bet for doing batch or historical processing through the renowned Hadoop framework so cloud-based analytics is the overwhelming practice. However, the emerging trend is to come with micro-scale clouds in between the ground-level sensors and the cyber-level cloud applications toward fog analytics. This specialized cloud, which is being formed out of networked and resource-intensive devices in that environment, removes the constricting stress on the traditional clouds. The proximate processing gets accomplished through these micro-clouds while the device data security and privacy are maintained. This kind of cloud-in-the-middle approach is capable of unearthing fresh IoT use cases. As any micro-cloud is very near the data-emitting sensors and sensor-attached assets, faster and cost-efficient processing and responses are being achieved.

2.1 It is all about the extreme and deeper connectivity

As the inventive paradigm of networked embedded devices expands into multiple business domains and industry verticals such as manufacturing facilities and floors, healthcare centers, retail stores, luxury hotels, spacious homes, energy grids, and transportation systems, there is a greater scope for deriving sophisticated applications not only for businesses but also for individual consumers. The world is becoming more and more connected. Recent devices include as standard connectivity features, and there is a wide range of a vast number of hitherto unconnected legacy devices. Furthermore, there is a wide range of resource-constrained devices ranging from heart rate monitors to temperature and humidity sensors, and enabling these to be integrated with other devices and web applications is a significant challenge. Thus, connectivity solutions and platforms are being brought in to enable every tangible device to be connected. Connectivity is required not only with adjacent devices in the vicinity but also with remotely held applications and data sources on the web/cloud.

2.2 The enormous volumes of IoT data

We have been investigating transaction systems extensively. IT infrastructures, platforms, and applications are designed to be appropriate for streamlining and speeding up transactions. However, with the faster penetration of devices and digitized entities, there is a relook. That is, operational systems are becoming more prevalent and prominent. In the impending IoT era, a sensor or smart device that is monitoring temperature, humidity, vibration, accelerations or numerous other variables could potentially generate data that need to be handled by back-end systems in some way every millisecond. For example, a typical Formula One car carries 150–300 sensors, and more controllers, sensors, and actuators are being continuously incorporated to achieve more automation. Today, all these sensors already capture data in milliseconds. Racecars generate 100–200 KB of data per second, amounting to several terabytes in a racing season. There are twin challenges for back-end systems. Storage concerns and real-time processing of data are equally important. Missing a few seconds of sensor data, or being unable to analyze it efficiently and rapidly, can lead to risks and, in some cases, to disasters.

2.3 Major IoT data types

There are three major data types that will be common to most IoT projects:

Measurement data: Sensors monitor and measure the various parameters of the environment as well as the states of physical, mechanical, electrical, and electronics systems. Heterogeneous and multiple sensors read and transmit data very frequently; therefore, with a larger number of sensors and frequent readings, the total data size is bound to grow exponentially. This is the crux of the IoT era. A particular company in the oil and gas industry is already dealing with more than 100 TB of such data per day.

Event data: Any status change, any break-in of the threshold value, any noteworthy incident or untoward accident, and any decision-enabling data are simply categorized as event data. With devices assisting people in their daily assignments and engagements, the number of these events is likely to shoot up. We have powerful simple and complex event processing engines in order to discover and disseminate knowledge out of event data.

Interaction and transaction data: With the extreme and deeper connectivity among devices, the quality and quantity of purpose-specific interactions between devices are going to be greater. Several devices with unique functionality can connect and collaborate to achieve composite functions. Transaction operations are also enabled in devices. Not only inter-device communication but also human-device communication is fairly happening.

Diagnostics data: The delectable advancements in the IoT domain have led to millions of networked embedded devices and smart objects, and information, transactional, analytical, and operational systems. There are online, off-premise, and on-demand applications, data sources, and services in plenty. The application portfolio is consistently on the rise for worldwide enterprises. There are software infrastructure solutions, middleware, databases, data virtualization and knowledge visualization platforms, and scores of automation tools. The health of each of these systems is very important for the success of any business transaction. Diagnostics data provide an insight into the overall health of a machine, system, or process. The data might show not only the overall health of a system but also whether the monitoring of that system is working effectively.

Precisely speaking, IoT data are going to be big and we have techniques and platforms for big data processing. However, the intriguing challenge is to do real-time processing of IoT big data. Researchers are working to unearth path-breaking algorithms to extract timely insights out of big data. Fog computing is one such concept prescribed as a viable and venerable answer for the impending data-driven challenges.

The IoT is turning out to be a primary enabler of the digital transformation of any kind of enterprising businesses. Companies are eagerly looking toward pioneering digital technologies to create and sustain their business competitiveness. The IoT and other digital technologies are helping companies to facilitate process enhancement, create newer business models, optimize the IT infrastructures, bring forth competent architectures, empower workforce efficiency and innovation, etc. The IoT closes the gap between the physical and cyber worlds, and helps to connect physical and digital environments. Data collected from connected devices are subjected to a variety of investigations to extract reliable insights.

3. The use cases of fog/edge computing

The rapid growth of personal, social, and professional devices in our daily environments has seeded this inimitable computing style.

Communication becomes wireless, sensors and devices are heterogeneous and large in number, geo-distribution becomes the new normal, interconnectivity and interactions among various participants emit a lot of data, etc. Massive amounts of data are being generated and gathered at the edges of networks.

Usually, these data are transported back to the cloud for storage and processing, which requires high bandwidth connectivity. In order to save network bandwidth, there is a valid proposition of using a moderately sized platform in between to do a kind of pre-processing in order to filter out the flabs. Differently enabled cameras, for example, generate images and videos that would aggregate easily in the range of terabytes. Instead of clogging expensive and scarce network bandwidths, a kind of fog/edge processing can be initiated to ease networks. That is, reasonably powerful devices in the environment that is being monitored can be individually or collectively leveraged to process cameras-emitted files in real-time. That is, the data gleaned can be subsequently segmented and shared to different nearby devices in order to do the distributed processing quickly. With more devices being added to mainstream computing and the amount of data getting stocked is growing exponentially, the distributed computing concept has soared in the recent past and is being touted as the best way forward for the data-centric world.

There are a number of convincing use cases for fog/edge computing. Fog devices locally collect, cleanse, store, process, and even analyze data in order to facilitate real-time analytics and informed decisions. Research papers describe how connected vehicles, smart grids, wireless sensor and actuator networks, etc. are more appropriate and relevant for people with the fast-moving fog computing paradigm. Smart building, manufacturing floors, smart traffic and retail, and smart cities are some of the often-cited domains wherein the raging fog idea chips in with real benefits. Augmented reality (AR), content delivery, and mobile data analytics are also very well documented as direct beneficiaries of fog computing. One use case for fog computing is a smart traffic light system, which can change its signals based on surveillance of incoming traffic to prevent accidents or reduce congestion. Data could also be sent to the cloud for longer-term analytics. Other use cases include: rail safety; power restoration from a smart grid network; and cybersecurity. There are connected cars (for vehicle-to-vehicle and vehicle-to-cloud communication), and smart city applications include intelligent lighting and smart parking meters.

3.1 Smart homes

A home security application is discussed in depth in a research paper. There is a myriad of home security products (smart locks, video/audio recorders, security sensors and monitors, alarms, presence, occupancy, and motion sensors, etc.). These are standalone solutions and due to disparate data transport protocols and data formats, these products do not interoperate with one another. However, fog computing has simplified the process of dynamically integrating these diverse security products in order to enhance the timeliness and trustworthiness of any security information. Fog computing platform is unique in that it can be flexibly deployed on a virtual machine or in a Docker container. Existing and new sensors and actuators are registered and connected with the fog platform, which ensures a seamless and spontaneous interoperation between different and distributed devices and machines to achieve the goal. This ad hoc collaboration capability senses any kinds of security threats and immediately stimulates the necessary countermeasures through connected actuators. Energy management, device clustering and coordination, ambient assisted living (AAL), activity recognition/context-awareness for formulating and firming up people-centric services, etc. are getting streamlined with the fog computing nuances.

3.2 Smart grids

A smart electric grid is an electricity distribution network with smart meters deployed at various locations to measure the real-time power consumption level. A centrally hosted SCADA server frequently gathers and analyzes status data to send out appropriate information to power grids so that they can adapt accordingly. If there is any palpable increment in power usage or any kind of emergency, this information will be instantaneously conveyed to the power grid for action. Employing fog computing, the centralized SCADA server can be supplemented by one or more decentralized microgrids. This salient setup improves scalability, cost-efficiency, security, and rapid response of the power system. This also helps to integrate distributed and different power generators (solar panels, wind farms, etc.) with the main power grid. Energy load-balancing applications may run on edge devices such as smart meters and microgrids. Based on energy demand, availability, and the lowest price, these devices automatically switch to alternative energies like solar and wind.

3.3 Smart vehicles

The fog concept can also be extended to vehicular networks. The fog nodes can be deployed along the roadside, and send and receive information to and from vehicles. Vehicles through their in-vehicle infotainment systems can interact with the roadside fog systems as well as with other vehicles on the road. Thus, this kind of ad hoc network leads to a variety of applications such as traffic light scheduling, congestion mitigation, precaution sharing, parking facility management, traffic information sharing, etc. A video camera that senses an ambulance's flashing lights can automatically change streetlights to open lanes for this vehicle to pass through traffic. Smart streetlights interact locally with sensors and detect the presence of pedestrians and bikers, and measure the distances and speeds of approaching vehicles.

3.4 Smarter security

Security and surveillance cameras are being fitted in different important junctions such as airports, nuclear installations, government offices, retail stores, etc. Furthermore, nowadays smartphones are embedded with powerful cameras to take selfies as well as produce photos of others. Still, as well as running images can be captured and communicated to nearby fog nodes as well as to faraway cloud nodes in order to readily process the photos and compare them with the face images of radicals, extremists, fundamentalists, terrorists, arsonists, trouble-makers, etc., in the already stored databases. Furthermore, through image processing and analytics, it is possible to extract useful information in the form of unusual gestures, movements, etc. All these data empower security and police officials to proceed in their investigations with clarity and confidence. Fig. 3 pictorially conveys how the fog cloud facilitates real-time sensor data processing and historical sensor data processing at nearby or faraway clouds (public, private, and hybrid) (Fig. 1).

3.5 Smart buildings

Like homes, office and corporate buildings are stuffed with a number of sensors for minute monitoring, precise measurement, and management. There is a school of thought that multiple sensor values, when blended, generate more accurate data. There are advanced sensor data fusion algorithms, and hence smart sensors and actuators work in tandem to automate and accelerate several manual tasks. For providing a seamless and smart experience to employees and visitors, the building automation domain is on a fast

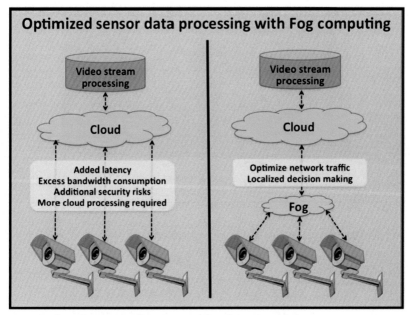

Fig. 1 The fog ensures zero latency toward real-time applications.

trajectory with a series of innovations and improvisations in the IT space. That is, computing becomes pervasive, communication is ambient, sensing is ubiquitous, actuation is intelligently accomplished, etc. The computer vision and perception topics are gathering momentum, knowledge engineering and enhancement are becoming common and cheap, and decision-enablement is gradually being perfected. The edge devices participating in and contributing to the edge cloud facilitate multiple things intelligently so that the strategic goal of building automation through networking and integration may be accomplished.

Traditional approach: Manual operation

Today, a medium-sized office building could have hundreds of sensors on its equipment. A great example is chillers, a product needed to cool a building. The product manufacturer (http://www.johnsoncontrols.com) monitors chillers remotely using predictive diagnostics to identify and solve issues before they become problems. The company uses internal operational data and historical records to plan machine maintenance more effectively, leading to better operational efficiency and decreasing energy usage, in addition to increasing reliability and equipment life span. Even better, the company has external data resources like weather patterns and grid demand costs to drive greater operational savings.

Chiller product.

There are several other industry verticals and business domains keen to gain immense benefits from the decisive and impressive advancements in the field of fog computing.

4. Fog and edge computing on digital twin

In many sectors like business, IT, industries, and also in power management wants foster collaboration for making any critical decision toward the progress. Cloud services, artificial intelligence (AI), and machine learning have been introduced by technical workers to integrate technologies and satisfy their requirements. When smart devices are introduced into the market to make something easy and connect anything anywhere, then the requirements toward technology increase in terms of performance, real-time response, and resilience and fault tolerance. Hence, to achieve these parameters, a new concept is evolved according to the basic "divide and conquer" approach.

As we discussed in Section 1, cloud computing provides centralized architecture and satisfies the above requirements, but some of the limitations such as round-trip time and heavy resources to approach cloud center such as bandwidth, gateways modulator require more which increase cost. To overcome these limitations, fog and edge computing were developed. As shown in Fig. 2, fog computing looks like cloud computing near smart devices, which are also known as IoT devices, and edge computing looks as if the cloud is integrated to those edge devices.

The diagram shows that he number of gateways is greater between IoT devices and the cloud, and the link represents the bandwidths (green = high bandwidth, black = medium bandwidth, and yellow = low

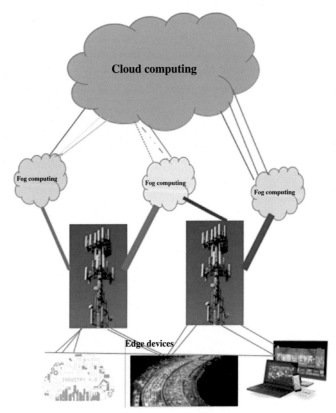

Fig. 2 Bandwidth utilization.

bandwidth) required for data transmission, which are also high. This concludes that if the distance between the devices for data transmission is less than better to choose fog/edge computing.

4.1 Fog computing

Fog computing is a distributed architecture which can provide services such as computing, storage, communication, and many others like cloud center. These resources are available near IoT/Arduino/edge devices. Fog computing overcomes the limitations of cloud computing, which are as follows.

4.1.1 Limitations of cloud computing
1. Response time delay.
2. Bandwidth utilization.

3. Connectivity.

4. Security.

Fog computing addresses these challenges by placing the required services near the edge device. As discussed by Bonomi [3], fog computing is a virtualized cloud center which collaborates with and distributes its service to any new technology or a standards to make the connectivity stronger and quicker.

Some of the characteristics of fog computing are as follows:

- Supports mobility.
- Scalability by using grid topology.
- Performance with better response time.
- Heterogeneity by supporting different devices.

The above-discussed features facilitate agility, proactiveness, and logistical data for artificial or machine learning devices.

The digital twin of any entity require enormous amounts of data, both historical and current, to predict and analyze the future consequences. Hence the internet and connectivity also play major roles in digital twin, and incorporating fog in digital twin makes the process robust.

4.2 Edge computing

Edge computing is a technology which enables data analyzing, storage, and utilization in edge devices. It allows collecting, aggregating, and communicating of the data along with providing intelligent logistics to give optimal solutions. Edge computing is a form of cloud computing and is also called micro-cloud services, which are integrated in edge devices.

A glance around at any devices used will find that they are are smart, quick, and robust, such as Google Home, Smart Home, smart home appliances, drones, Amazon Dash buttons, and many more. A research group estimated that the drastic growth of these smart devices may reach around 75.44 billion by 2025.

One of the backbone technologies is edge computing; it is a gateway between edge devices to connect to the internet with minor latency which is suitable for real-time applications.

Edge computing is capable of managing applications, providing security, connectivity, and also scalability. It involves not only the software components but also hardware which interacts with the user and makes faster decisions.

Some of the vendors for edge computing [4] in markets are:

1. FUJITSU IoT Solution: provides edge computing to bridge the gap between operational and tradition devices in an industrial context.
2. FogHorn: provides edge computing for IoT.
3. Saguna: provides a multiple-access edge cloud.
4. ClearBlade: orchestrates multiple layer and edge computing devices.

Many more, including Google, Cisco, and Microsoft, are providing edge gateways and analytics.

The role of edge computing is vital on digital twin, as it:

- provides application logic to the components in a model;
- filters the streaming data before reaching the destination;
- provides diversified data to the components;
- enables synchronization between devices or components;
- maintains the statistical information of devices;
- offers an easy access point; and
- enables peer-to-peer connectivity through mesh topology.

5. Facets of digital twin

Digital twin is a virtual representation of physical devices used in most IoT devices to build and engineer with a defect-proof model. Technically it is a concept to simulate each and every dynamical movement of data from electrons to elements. It not only simulates but also gives status and condition according to the load and age of the model.

Hence collaboration of digital devices with edge/fog computing with artificial intelligence makes models robust and perfect.

So the facet of digital twin is not only the simulated model but also dynamic model. Hence, it is a device that serves the following purposes, among many others:

- prediction
- analysis
- simulation
- projection
- demonstration
- collaboration (Fig. 3).

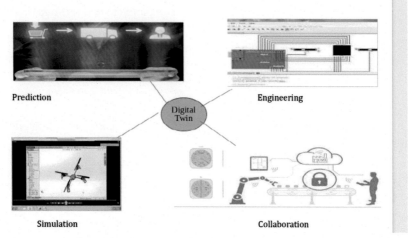

Fig. 3 Facets of digital twin.

(a) *Prediction*: Digital twin is a model which can take a data set of different patterns from machine data and experiment with data fusion to predict the possibilities.

(b) *Analysis*: Analyzing the gathered data from all the relevant component through any of the data science techniques can give an appropriate input to predict.

(c) *Simulation*: Digital twin is not just a simulator to simulate the system, but also provides ideas about upcoming challenges and adopts the new properties to rebuild the model.

(d) *Projection*: Projection is basically used as a business process to integrate the requirements and identify a solution if one is needed.

(e) *Collaboration*: Digital twin can collaborate with any technologies to bridge the gap between past and future. It can integrate with cloud/fog/edge or any AI to improve the efficiency.

(f) *Demonstration*: A sample model can be demonstrated and the available data experimented with for refinement.

Fig. 4 visualizes the collaboration feature of digital twin.

Those properties of digital twin prove that it can integrate with any standards and technology to build a fault-proof model.

6. Collaboration with fog computing

Digital twin is an extended feature of the IoT which provides deeper insights into any model to automate and re-engineer it to fix the bugs and make it agile.

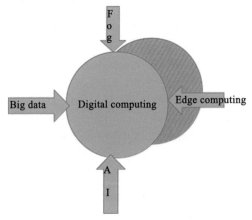

Fig. 4 Digital twin collaboration.

Further to the discussion of fog computing and its capabilities, we shall now consider the scope of collaboration of fog computing with digital twin by discussing the step to build digital twin. The role of fog computing bridges the gap between digital twin and the internet.

Step 1: Construct the simulation model of physical model components.

Step 2: Store all the training data between components. (*Storage*)

Step 3: Sense all the external factor impacts on the model. (*Connectivity with low latency*)

Step 4: Compute the logic (algorithms) for all "if else" statements. (*Computing power*)

Step 5: Identify the point of failure and rebind quickly. (*Server/traditional structure*)

Step 6: Provide for scalability through integration or coupling. (*Standards*)

Step 7: Finally, resultant value impacts on markets. (*Global view*)

In the above-discussed steps, the highlighted parameters are the external components or the facilities required by digital twin. Fog computing not only provides the cloud services such as computing, storage, and infrastructure but also concentrates on performance and response time, which are prime factors in real-time application. As discussed in "A Fog Computing and Cloudlet Based Augmented Reality System for the Industry 4.0 Shipyard" [5], fog computing addresses the challenges such as transmission delay and network traffic latency, which are high in cloud and progressively less in fog computing.

7. Collaboration with edge computing

7.1 Edge computing

Edge computing is a distributed computing paradigm which brings data storage and computation or processing closer to the location where it is needed to improve response times and save bandwidth.

By pushing computation to the edge of the network, it is possible to analyze data in real-time which is crucial in industries such as manufacturing, healthcare, telecommunication, finance, etc. Edge computing is a mesh network of micro-data centers that process or store critical data locally and push all the received data to central data centers or cloud storage repositories (Fig. 5).

An ideal situation for edge computing deployment would be in circumstances where IoT devices have poor network connectivity and also as it is not very efficient for IoT devices to be always connected to the cloud. Edge computing can be used in areas such as financial services and manufacturing, which are sensitive to latency. Latency of even milliseconds in processing of information may be untenable for such applications. Edge computing reduces latency as data need not be transferred to the cloud or data center over the network for processing. The aim of edge computing is to push computation to the edge of the network away from data centers, exploiting the capabilities of smart objects, cell phones, and network gateways to provide services and processing on behalf of the cloud.

An example of edge computing deployment would be an oil rig in the ocean that has many sensors generating massive amounts of data which perhaps confirm the proper functioning of the rig. However, most of the data generated by those sensors may be inconsequential and hence do not have to be sent across the network as soon as the data are produced. Thus, the local

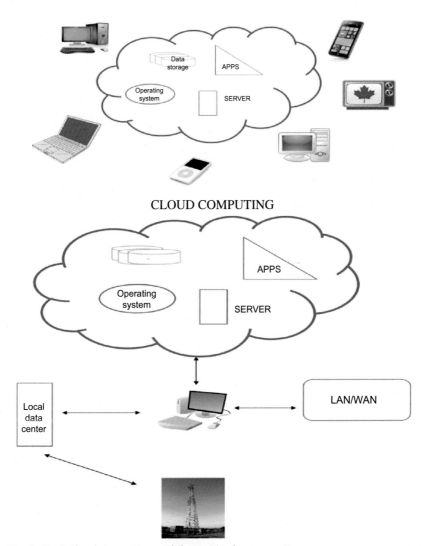

Fig. 5 (Top) Cloud computing and (bottom) edge computing.

data computing system compiles the data and generates daily reports, and sends the data to the cloud or a central data center for storage, thus sending only important data and minimizing the amount of data transferred across the network, saving network bandwidth and also reducing latency.

Another example which explains the benefits of edge computing is the usage of web browsers by internet users. Thousands of people are connected to the internet at any given point of time. One of the most widely used applications by these users is the web browser. Browsers are used to watch videos, for research, etc., and are operating on more and more devices such as cell phones, set-top boxes, and stick PCs. The usability and display speed of the web browser depends upon the performance of the device. If the device does not perform very efficiently, the latency increases which can be very stressful for the user. Edge computing can be deployed in order to overcome the latency thus created. An edge server can be placed between the cloud and physically closer to the user. The Nippon Telegraph and Telephone Public Corporation (NTT) has developed the NTT web browser which can offload part of the workload of the devices to the edge servers. The processing is done by the edge servers, freeing the device of the task, thus reducing latency. The content is displayed at a much faster rate than through a standard web browser.

Edge computing dates back to the 1990s and is still considered a new paradigm, despite its age. Edge is a new buzzword which simply means processing and analyzing data along the edge of a network, nearer to the point of data collection, so that data become actionable. The objective of edge computing is to solve the proximity problem, thus solving the latency problem. Since edge computing does not depend only on the cloud for processing, outage reduction and intermittent connectivity can be improved. In addition, by ensuring reliable operations in remote locations, unplanned downtime as well as server downtime can be avoided.

7.2 Collaboration

We can maximize the benefits of edge computing by combining it with other technologies. One such collaboration by which businesses can derive advantage involves combining it with Digital Twin technology. Edge computing has the potential to minimize risk in real-time whereas Digital Twin can predict future events accurately, which can benefit businesses to a great extent [6].

Digital Twin is a virtual or digitized model of a service, product, or a process or any IoT. More than one digital twin can be developed for any particular object based on its industrial context. It is a virtual representation of the components and dynamics of the real-world entity which facilitates operations of the product as well. Processing can be simulated using digital twin which can accurately predict the product's behavior under different conditions such as weather changes or as the product ages. This will help in making required changes in the product and making it a better one [5].

Digital Twin provide significant advantages in a cloud-based environment. What if we deployed a digital twin along the edge of a network? Edge-deployed Digital Twin gives way to new autonomous systems—real-time artificial intelligent systems based on self-learning. By developing a virtual model of the physical asset, greater flexibility can be achieved to define, evolve, build, and leverage twins for real-time IoT.

Digital Twin resides in and is maintained in the cloud, fed with data from its environment through devices, sensors, or simulations. Digital twins are used to understand past and present operations, and to predict future events by leveraging with machine learning approaches to build a better system by detecting anomalies and forecasting failures. Edge computing provides advanced processing and cloud-based analytics that enable faster and localized decision making at the edge [1]. Deploying the digital twin at the edge will make it possible to build smarter applications. Moving the digital twin to the edge of the network facilitates smarter applications for the following reasons:

- Reduction in latency: In a cloud-based digital twin, the delay caused by cloud access is not acceptable. Edge-deployed digital twin reduces the latency, and applications that require sub-second latencies can be derived.

- Analytics generated by the digital twin can guide local control and vice versa. An anomaly that could cause a problem in future can be fixed without human intervention.

- Using machine learning approach on streaming data, digital twin can evolve much faster and acquire the ability to self-learn.

- Edge computing technology implements decentralization of data, and distribution in the network; therefore, it will not be possible for hackers to corrupt data. In addition, transfer of sensitive information over networks will be minimized and thus data security will be enhanced. Data encoding and implementing virtual private networks become have very important with the growth of edge computing technology [3].

- Edge computing also facilitates scaling of IoT networks much faster and as required [1].

Edge-deployed digital twin also provides business benefits such as the following:

— Because cloud storage and analysis can be costly. Edge-deployed digital twin reduces cloud hosting costs as all the data need not be sent to the cloud.

— As the data processing is performed at the edge of the network, the volume of data that has to be transmitted to the cloud reduces.

- Sensitive data need not be sent to the cloud.
- Analytics can be performed even when the digital twin is disconnected. By merging these two emerging technologies, we can develop devices which can exhibit intelligence and possess immense decision-making capabilities without human intervention. We can create a future which we have never imagined before. Many systems and products combining these two technologies have been developed and deployed in various areas of business which impart both tangible and intangible benefits to the organization. The following section discusses some of the use cases which substantiate the above statement.

8. Use cases of digital twin collaboration with fog

8.1 Drones in agricultural fields

A drone is a flying agent to perform a specific task in any given area. Some of the experience shared by the students in IoT lab to design the drone fails successively in building process due to the imbalance in propeller and after they use the digital twin to simulate the drone and successfully built after simulating by providing proper parameter. This kind of experiment takes place in many areas to reap the benefits of IoT devices. At the same time, many new thoughts arrive as the requirement increases lead to the failure of existing IoT devices to support. Hence digital twin bridges the gap and collaborates with the internet to provide the service on demand with micro-latency through fog computing. As shown in Fig. 6, Skyx's

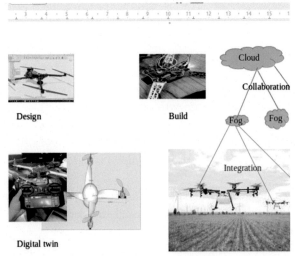

Fig. 6 Auto-piloting in agricultural fields.

developed a drone for auto-piloting in agricultural fields to manage and monitor crops development; this was also suitable for real-time controlling.

8.2 Automotive industry

The automotive industry has introduced electric cars into the market in recent years. There has been a growing demand for these electric cars which were developed with a benevolent intent of saving the environment. Though these cars are currently manually driven, the industry is aiming to convert these manually driven cars to self-driven bots. Edge computing and simulation technologies can help bridge the gap and make self-driven cars a reality [6]. In cloud-based technology, data processing and storage are done remotely, whereas in edge computing, data are processed at a location which is in close physical proximity to the data source, facilitating transfer of data at a faster rate. Using this technology, cars could make decisions such as when to apply a brake, start driving, make turns, increase speed, etc. much faster [6]. By combining this with digital twin technology and simulation, future events can be predicted which can help the vehicles to navigate better and avoid collisions by making appropriate decisions whenever the need arises.

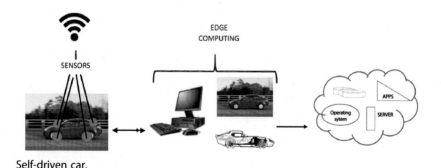

Self-driven car.

By enhancing and implementing this technology in the automotive industry, development costs can be reduced, efficiency improved, and sustainability can be enhanced by automobile manufacturers.

8.3 Windmills

Consider the hypothetical situation of a windmill where a hierarchy of digital twins can be used. In this example, the digital twins are organized in a hierarchy where the low-level digital twin represents individual components and the higher-level digital twin represents subsystems which control these devices. Twins at various levels send the messages downward to the lower levels for controlling, as a result of which signals are generated that will in turn be sent to the devices. The application logic has to be divided between the cloud and the edge. Digital twins enable to successfully migrate low-level event handling operations to the edge and higher-level digital twins can operate in the cloud or wherever the computing resources are located (Fig. 7).

The use of digital twin technology does not require all device-specific operations to be migrated to the edge of the network. Instead, the low-level twin can the functionalities of the device directly, and the high-level twin implements a machine learning approach and performs predictive analytics based on data received by the low-level twin. Thus, it is a good idea to move the low-level twin to the edge and hence reduced response time and processing can be achieved without any interruption. The high-level twin can reside wherever the computing resources are located or in the cloud for the predictive analytics algorithm to be executed [2].

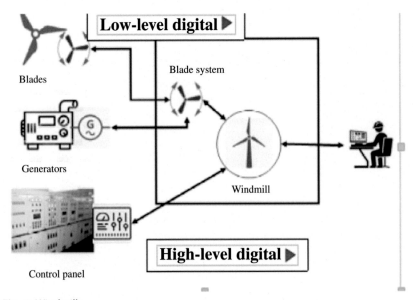

Fig. 7 Windmill.

8.4 Workplaces

The workplace environment can be checked in real-time by deploying edge computing with digital twin technology. By creating virtual representations of buildings and office spaces and integrating these with edge computing, the environment can be monitored in real-time, potential risks and dangers can be avoided, and suitable measures can be taken to avoid accidents and disasters.

A few use cases were discussed in the above section, but this is not all.

Although edge computing and Digital Twin technology are being implemented extensively in various sectors of business and industry, the combination of these technologies is also being deployed in today's business which can provide real-time, intelligent solutions for organizations. We are heading toward a future which we have never experienced before.

9. Benefits

Collective data analysis to compare the performance on digital twin between cloud, fog, and edge computing was carried out with a list of parameters including:

- speed
- bandwidth
- response time
- latency
- throughput.

As shown in Fig. 8, drastic improvement in edge computing with respect to the above-listed network parameters signifies the integration of digital twin with fog/edge computing.

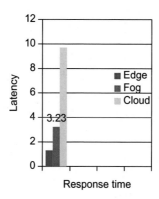

Fig. 8 Response time.

10. Conclusion

Digital twin technology empowers edge/fog devices to be intelligent in their actions and reactions. As the future beckons for edge/fog clouds for realizing and running real-time and insight-driven applications and services, the role and responsibility of having and using digital twins for all the participating physical twins are bound to become more significant. Furthermore, the analytical capabilities of both physical and digital twins are useful in making edge/fog devices (physical twins) adaptive, adjustive, and autonomous. This chapter focused on describing how the digital representation of edge devices is ultimately helping to empower them to be intelligent in decisions, deals, and deeds.

References

[1] https://www.iotforall.com/fog-vs-edge-computing-do-differences-matter/.
[2] https://docs.oracle.com/en/cloud/paas/iot-cloud/iotgs/iot-digital-twin-framework. html.
[3] Bonomi F, Milito R, Zhu J, Sateesh Addepalli Cisco Systems Inc: *Fog Computing and Its Role in the Internet of Things,* 2012. 170 W Tasman Dr. San Jose, CA 95134, USA.
[4] https://www.zdnet.com/article/10-edge-computing-vendors-to-watch.
[5] https://www.ncbi.nlm.nih.gov/pmc/articles/PMC6022113/.
[6] https://www.electronicdesign.com/communications/understanding-wireless-routing-iot-networks.

Further reading

[7] https://www.forbes.com/sites/ralphjennings/2019/08/15/taiwan-will-easily-overcome-chinas-ban-on-82000-tourists-per-month/#6a208ec55781.
[8] https://cloudblogs.microsoft.com/industry-blog/manufacturing/2018/08/20/the-cloud-enables-next-generation-digital-twin/.
[9] https://agfundernews.com/skyx-crop-spraying-drone-raises-seed.html.
[10] Alam KM, Saini M, El Saddik A, et al: *IEEE Access* 3:343–357, 2015.
[11] kazi Masudul Alam and Abdulmotaleb el Saddik: C2PS: a digital twin architecture reference model for the cloud-based cyber-physical systems, *IEEE Access* 5:2050–2061, 2017.
[12] https://www.eclipse.org/ditto/intro-digitaltwins.html.
[13] https://www.logistics.dhl/content/dam/dhl/global/core/documents/pdf/glo-core-digital-twins-in-logistics.pdf.

About the authors

J. Pushpa is a research scholar in VTU Belgaum. Her areas of specialization are software defined networking, edge computing, and fog computing. She has gained experience in the IT industry and in teaching organizations, and is currently working as an Assistant Professor in Jain University, Bangalore.

S.A. Kalyani is serving as an Assistant Professor in East West College and has enormous knowledge in the area of artificial intelligence and working on data science.

CHAPTER FOUR

The industry use cases for the Digital Twin idea

Peter Augustine

Department of Computer Science, CHRIST (Deemed to be University), Bangalore, India

Contents

Abstract

Digital Twin Technology has taken the place in top 10 strategic technology trends in 2017 termed by Gartner Inc. Digital Twin concept brings out the virtual depiction or the digital representation of the real world equipment, device or system whereas the real world and the virtual world gets the highest synchronization. The digital representation of the complete life cycle of a product from its design phase to the maintenance phase will give the prophetic analysis of the problems to the business.

This greatest advantage of foreseeing problems in the development of a device will give early warnings, foil downtime, cultivate novel prospects and inventing enhanced devices or gadgets for the later use at the lesser expense by means of digital representations. Indeed, these will devise a larger influence on conveying superior consumer feeling also in the enterprise. The emerging trends such as Artificial Intelligence, Machine Learning, Deep Learning, Internet of Things and Big Data used in Industry 4.0 play a vital role in Digital Twin and they are mostly adopted in the world of manufacturing, Industrial Internet of Things, and automobile business world. The penetration, wide coverage and the advancement of the Internet of Things in real-world have elevated the power of Digital Twins more economical and reachable for the world of various businesses.

1. *Manufacturing*: Digital Twin has brought out the change in the existing manner of the manufacturing segment. Digital Twins have a substantial influence on the design of products and their manufacturing and maintenance. Because of its influence the manufacturing more competent and augmented while dropping throughput times.

2. *Industrial IoT*: Integrating digital twin with industrial firms will facilitate the activities such as monitoring, tracking and controlling industrial systems in digital means. We can potentially experience the power of digital twin since it captures environmental data such as locality, settings of the devices, financial frameworks, etc., other than the operational data, which benefits in foreseeing the forthcoming operations and incongruities.

3. *Healthcare*: Since the healthcare sector demands higher accuracy in diagnosis and treatment, with the important data from IoT, digital twins can play a vital role by reducing the expense for the patient, precautionary alerts to avoid health deterioration and giving tailored health support system. This will be great support especially in developing countries like India.

4. *Smart cities*: Digital Twin coupled with IoT data can augment the efficient planning of the smart city and execution of its building by supplementing financial progress, effectual administration of resources, lessening of environmental impression and escalate the complete worth of a resident's life. The digital twin prototypical can aid city organizers and legislators in the smart city planning by retrieving the visions from numerous sensor networks and smart systems. The information received from the digital twins supports them in reaching well-versed choices concerning the future as well.

5. *Automobile*: Automobile industry can get voluminous benefits out of Digital Twins for producing the simulated framework of a coupled vehicle. It retrieves the behavioral and functional information of the vehicle and services in examining the inclusive performance efficiency of the vehicle as well as the features connected along with it. Digital Twin also supports in supplying a justly enhance support and service for the consumers.

6. *Retail*: Alluring client satisfaction is a fundamental factor in the merchandising world. Digital twin employment can play a key role in supplementing the retail customer experience by forming virtual twins for customers and modeling fashions for them on it. Digital Twins also supports enhanced planning of stock maintenance, safekeeping procedures, and human resource administration in an augmented means.

The word 'digital twin' was, to begin with, devised in item development administration in 2003 when computerized illustrations of physical items were still in their infancy. With the development of computational power and the devices intertwined with the Internet, Digital Twins are presently picking up footing over businesses of various fields. They were as of late named one of the greatest 10 significant innovation patterns in 2018 by Gartner.

1. Manufacturing

In the face of profound transformation in the manufacturing world and the essential to capitalize on Industry 4.0 bonanza, the digitization of equipment, processes, and products is gaining more relevance. Although it has been around for centuries, the notion of a physical asset's digital replica is about to enter a new age. The reduction in the cost of IoT sensors and the advancement in their functionalities and the equipment coupled with those IoT devices for manufacturing are the major causes which force the manufacturing units to use the digital twins for the better quality products, better usage and well managing the human and hardware resources.

From the time of inception, the digital twin has taken dynamic evolution in the manufacturing world by influencing the product lifecycle management. Now the evolving features of the digital twin can be adopted in different and vibrant perspectives in manufacturing especially with respect to the asset and the product. In the manufacturing industry when equipment such as a physical asset is digitally represented and they are connected based on their functionalities, the equipment becomes the socket of smart factory and the supply chain. This smart digital representation can give and receive data to and fro from all levels of its connected equipment involving in the manufacturing process. Meanwhile, the digital representation of the equipment can also be read by the other vendors involving in the business, the operators, maintenance engineers and the regulatory bodies too.

When a product is featured with digital twins, there are more benefits such as having control in every moment over the complete life cycle of the product and if any abnormality arises with respect to the predesigned model, there is a complete control to alter the process. In the meantime, it is challenged that application of digital twins in small-scale industries may be an encumbrance for a short while with respect to cost and implementation intricacies but indeed for a long run, it would definitely be a gift of benefits in all aspects (Fig. 1).

Fig. 1 Digital Twin in manufacturing. *Source: https://internetofbusiness.com/half-of-busi nesses-with-iot-projects-planning-to-use-digital-twin/.*

1.1 The business benefits of Digital Twin in manufacturing

Organizations willing to be front-runners rather than laggards need to cascade these priorities down business models, processes, and working models and reshape them accordingly.

(a) *Connecting places of processes*

In the real world, the manufacturing processes are taking place in different sites of distances where a high degree of physical connectivity is demanded all the time. Indeed digital twins undoubtedly yield in lessening the troubles in connecting workforces of different places in the manufacturing industries. The digital replica of the industry can help to devise new models to augment the supply chain for the effective time management between the places of operations such as customization of the workforces, movement of the spare parts and rescheduling the process to enhance the efficiency.

(b) *Optimizing services*

Maintenance plays a vital role in any business and so in the manufacturing industry too. From the time of inception, the product gets created through various processes and still, it demands to care even while it is in the hands of the customer. The model to enable efficiency in handling third-party vendors and customers with respect to time, pricing, operating performance, eradicating latency in receiving and giving products, best customer service and handling the warranties can be achieved using digital twin. This will help in proliferating business opportunities.

(c) *Enabled expert system*

As in the case of big data, the dynamic growth and development in the technologies involving in the manufacturing digital devices cause the production of a huge amount of data. These data interact with each other and cause for the immediate response in the devices, processes, and people. Digital twins in such a modern scenario of manufacturing can lead to an expert system which replicates an efficient decision support system responding positively to the environment. This expert intelligent system can suggest the innovative solution for immediate realigning of the human resources across various units or alteration in the behavior of assets. The environmental factors such as demand in the market and the external parameters affecting the demand and supply can also be addressed by this immediate response system and it can certainly increase the share value of the company too.

(d) *Standardizing process, places, people and machinery*

The amalgamation of various electronic devices connected to the internet world demands standardization of the devices, the communication standards, and the data sharing mechanisms. Digital twins replicating the machinery producing data and involving processes and people also is not exempted from this standardization. But it is an added advantage if prototypical standardization is ensured in terms of interoperability and security issues in the internet world.

(e) *Feedback and reengineering*

In the current customer based business model as the customer is the boss, it is more appropriate to respond in the proactive and reactive ways based on the customer reviews and the satisfaction surveys. It is really a masterpiece in improving a company's performance and the stock value in the market and finds the right place of its own in the manufacturing world. In earlier days user design alone took the prominent place in the design phase whereas nowadays the user experience has also coupled with it. It shows the need for the user experience by getting the authorized feedbacks and respond positively by adopting in the processes and the work style of the employees instantaneously. Digital twins are the right and appropriate choice in this scenario and the high demanding customer world to sustain the business by adopting the changes in the environments and fixing the problems predicted in the right place.

(f) *The evolution of maintenance*

Time-based maintenance has been more or less not required in favor of preventive, rule-based strategies. In this respect, most common industry benchmarks show up to be constrained as a result of the little sum of information that can be prepared and the failure to analyze inconsistencies other than limits-breaking. In spite of holding on lawful commitments around preventive approaches, the combined appearance of cheap sensors and fake insights is bringing support one step advance: real-time field information can be utilized to bolster machine learning calculations competent of recognizing abnormal patterns and anticipating disappointments well sometime recently they really happen. Typically anticipated to convey critical benefits in guaranteeing tall asset accessibility and optimizing benefit endeavors, especially with prescriptive upkeep, which is ostensibly the following developmental organize. Not as it were will the framework recommend parts for substitution or repair. It will issue a work arrange at that point set up and direct the related workflow.

The following are the two eye-catching cases of manufacturing companies succeeding with this digital twin innovation:

Kaeser compressors

The manufacturer of compressor items from the United States, Kaeser utilized Digital Twins to go from merely offering an item to offering a benefit. Rather than introducing hardware at a customer's location and taking off operation to the client, Kaeser keeps up the resource all through its lifecycle and charges expenses based on discussing utilization instead of a settled rate. A Digital Twin arrange empowers the company to screen the condition of its hardware around the clock and degree client discuss utilization. Real-time resource information makes a difference Kaeser with guaranteed hardware uptime and charging the exact sum of cash for each billing cycle. To date, the company has cut product costs by 30% and onboard 50% of major sellers utilizing digital twins.

Stara tractors

Brazil-based tractor producer Stara employments digital twins to modernize cultivating. By outfitting its tractors with IoT sensors, the company can increase the hardware execution. With real-time input of the functionalities of the tractors, Stara can proactively anticipate gear glitches and progress resource uptime. The company has moreover utilized Digital Twins to form modern trade models. With a riches of IoT sensor information, Stara propelled a beneficial modern benefit that gives agriculturists with real-time knowledge specifying the ideal conditions for planting crops and making strides cultivate abdicate. Farmers have decreased seed utilize by 21% and fertilizer utilize by 19% which is highly appreciated to Stara's direction.

1.2 Challenges for Digital Twins in the manufacturing industry

In spite of the fact that numerous driving fabricating businesses are contributing to advanced change, the way can be cleared with deterrents. A handful of vital needs must be tended to undertake to dodge the risks of such a journey:

1. Reframe the world view in the modern era and clearly distinguish the benefits to be achieved. Translate operational technology (OT) and information technology (IT) merging into an innovation adoption or refresh roadmap.
2. Define the organizational structure that best fits the unused set of capabilities that hide inside the workforce and the collaborators and the structure which gives a clear vision on the modules, requirements, and collaboration models.
3. Grow incrementally by beginning with little, reliable trial circumstances and, once it demonstrates effective, scaling it up and out.

2. Industrial IoT

Digital Twin can provide a large-scale perception on most dimensions when it is added to the IoT. This recommends it arranges various parts of an IoT gadget. The Digital Twin makes working with IoT resources exceptionally simple by connecting the physical model and the digital interface. A Digital Twin is simply the digital representation of the object and connecting with the real world object to access and work with various capacities and highlights of that object.

The significant advantage of Digital Twins in the IoT is that it is not necessary to stress over merging with the resources for continuous communication of any information. Relatively, IoT can just send requests to these physically distributed IoT resources in a safe sandbox in the cloud, which

Fig. 2 Digital Twin in IIoT. *Source: https:/solutionsreview.com/business-process-manage ment/gartner-survey-reveals-increased-iot-implementations-and-use-of-digital-twins-in-2018/.*

has been working with the Digital Twins which gives the scenario of IoT and Digital Twins are working together in the same workspace (Fig. 2).

This sandbox approach lessens the security threats as the applications are not transferred as digital twins does not require so, though they are in the cloud. It has been observed that the development expenses are declining and it denotes that the IoT applications can be developed faster. As a result, Digital Twins in the cloud can enable the opportunities to open up several novel volumes and provisions in IoT.

Digital Twins can support in the following ways while deploying IoT:

(a) *Predict the future*

Since the real-time massive information gathered from the IoT sensors can be realized in the Digital Twin, it is easy to predict the life span of the devices, their working conditions and change in their functionalities, because of the data flowing from the environment also.

(b) *Proliferation of precision*

It is not difficult for businesses to connect and communicate between data as in the case of big data since they arise from more connected gadgets and applications and it would be easier to acquire a better view of the complete condition of the Industrial devices. The amalgamation of the machine learning algorithms along with these interweaving data of behavioral patterns of the devices can help Digital Twins to understand precisely how a device may function in a particular scenario to soften potential failures in the devices before it takes place.

(c) *Producing complex Digital Twins*

Indeed the functionalities of Digital Twins support the software applications to connect with remote devices in real time easily and retrieve data from it or instruct the device to work in the desired manner. But this limited real-time data may not be enough to have complete control over the devices whereas the behavioral pattern of the past and the future predicted data are also mandated to enhance the precision. The twining of the behavioral functionalities with real-time digital twins can lead to the next level of success.

(d) *Lessen expenses*

Businesses always work with the concept of reducing costs and increasing the benefits. In the field of IoT also the companies evaluate the process of deploying the devices and testing new, rationalized solutions to reduce costs and accordingly increase revenue margins. Digital Twin allows the companies to test this without making real-time extensive physical deviations to the complete product. At a broader professional level, retail stores can adopt Digital Twin technology to imitate their store and, using sensors, track consumer activities with respect to purchase and reduce these expenses. In reality, Digital Twins helps to sustain a strategic distance from faults and differentiate accomplishment.

(e) *Evading failure*

Avoiding failure may look apparent in a real world scenario, but industries need to pay great value to avoid costly breakdowns or errors. Digital twins empower the architects to discover numerous potentials so they can deliver a commendation around the durability or consistency of an asset with a greater level of assurance. Industries that are willing to progress uptime and rise manufacture can leverage digital twins to do so extra rapidly.

By progressing from a single resource view to a bigger resource populace, industries can open novel probabilities to expand accomplishments. For instance, an engine, transmission, and brake mechanism may all have distinct Digital Twins, though must interconnect with one another just as much as the real physical engine, transmission, and brake mechanism to render deeper insights.

2.1 Use cases for Digital Twins in the IoT

The arenas of usage for Digital Twins are composite and not kept to a specific business or region. They can be employed for an extensive possibility of circumstances. There are Digital Twins that merely model and depict to a single sensor inside a gadget in one side. On the other side, there are Digital Twins that replicate various parts of an entire place of buildings, containing vitality, utilization, topology, and much more from there.

Digital Twins can, for instance, be deployed in the circle of the Industrial Internet of Things (IoT). Take the assembling procedure as a general model. By outfitting machines with sensors, it is easy to gather a collection of operational information. It is not merely identifying with the behavior of the machine itself also in addition to the environments in the processing plant involved. By utilizing a Digital Twin, we can connect and inspect these informational collections just as replication of production procedures in the virtual world. After some time, if deviations in execution become indistinct, producers can make a move to improve their production techniques.

There are use cases outside production and Industry 4.0, for example, a Digital Twin put to use with regards to the associated constructions. Here one can have a simulation of how construction is utilized, in light of recorded or relative information, and assess changes in the construction's plan. In this scenario, Digital Twin can point out areas that are exhausting resources or are utilized just once in a while.

Utilization-based insurance is another conceivable use case. Rather than conveying a costly telematics unit on each new client's vehicle, the application can now essentially be deployed in the cloud. The Digital Twin can be utilized to cost the driver's individual driving score continuously.

2.2 GE Digital Twin

There will be symptoms which reveal that the device is going to end with serious issues. Digital Twins is used by GE to observe and analyze the preliminary unstable functions of the devices. Even though there are anomaly detection frameworks are available already, creating such frameworks with digital twin can supplement to predict these abnormal functions early and enable the system for appropriate maintenance schedule in advance. The inputs flowing from the behavioral pattern of the assets, processes, systems and the corresponding complete domain knowledge along with the artificial intelligence incorporation will create an expert digital twin representation to work with. In GE, the usual anomaly detection predicts with the time slot 20–30 days advance has changed to 60 days advance while Digital Twin is adopted. This will definitely help in planning the corrective measures well in advance with all possible proper inputs. The Digital Twin is also used to forecast the lasting life of a turbine blade on a particular aircraft engine with excessive precision.

Industrial facilities are being altered by understanding the prospective consequences of specific assets. GE has also formed Digital Twins for the enterprises that simulate comprehensive, composite systems communications, which simulate numerous circumstances of the future and regulate best possible significant performance parameters for circumstances with maximum possibility. By leveraging enormous data sources for weather, performance, and operations, these simulations exhibit probable situations that might influence an enterprise.

2.2.1 Digital Twin ecosystem

As Digital Twin is capable of having complete 360-degree connectivity, the team in GE comprises of all the people in a specific domain like experts of such domain, designers of a model, data analysts and scientists, and business innovators to build a digital twin. It connects all these people of the broader community to retrieve maximum understandings and innovative ideas. Since GE digital twin framework has been built on Predix, the digital twin becomes accessible to all GE clients and business partners. The GE industry runs on the Predix platform which is a scalable, asset-centric data foundation. Predix works as a complete and safe application platform which helps to run, scale, and extend digital industrial solutions.

The GE partners are also now coming up with the Digital Twin ecosystem. Infosys is one of the GE Partners who adopted the digital twin approach in 2015 itself. The primary focus of this early stage tie-up was to enable early cautions and failure forecasts for different major components like landing gear in a flight by developing a Digital Twin and it is the world's first Digital Twin for this purpose.

When the insights received from the engine, airframe, and further systems are combined with Digital Twin, it can lead to customize a wider and advanced Digital Twin of the complete aircraft. This comprises of fuel efficiency, maximizing security, appropriateness, optimizing fuel consumption with operative quid pro quo among these components and the enterprise simulation. The elasticity advantage of digital twin ecosystem by accommodating the various dimensions can transform the industrial services along with the power of accurate prediction of the future.

Since the way, the twins are in communication and constantly learning from each other, the knowledge of machines communicating, manipulating the information, and making selections with each other will be transformative. In this way, digital frameworks are operated and administered within the future. There may be over 50 billion devices associated with each other through IoT along with over 7 billion internet users between the years 2020 and 2030.

3. Healthcare

The seventh crewed assignment in the Apollo space endeavors launched Apollo 13 shuttle in the month of April in 1970. But unfortunately one of their oxygen container labeled number two in the processing module exploded before the completion of the mission of the space travelers Jim Lovell, John Swigert, and Fred Haise. They were battling for life and there was a cry for life from Swigert to Houston. There was an emergency call from 200,000 miles away to safeguard the space travelers in the mission to bring them domestic securely. When the world was in great anxiety to get the travelers safely to the earth, the shuttle controlling team in NASA in Houston started working on this greatest challenge instantaneously.

NASA handled the malfunction of the shuttle by keeping the resembling framework of the Apollo 13 on the base as a real world model of the shuttle and its machinery. This helped the engineers in Houston on the ground to show and examine conceivable arrangements, reconstruction the settings coupled with Apollo 13. When the accumulated quantity of carbon dioxide in the lunar module of the Apollo 13 shuttle increased to dangerous levels, NASA engineers made an adapter with better-quality settings using all sort of weird and random parts, like a flight manual cover, parts of the flight suit, and socks. They instructed the space travelers how to construct it with resources accessible within the spaceship. Simultaneously, space travelers on the ground base ran a trial at Houston and Kennedy Space Center to test methods for getting the team of Apollo 13 returned to the earth safely which they inevitably are successful in, 4 days afterward the mishap.

Undeniably, the modern era of innovation has begun since 1970. In the case of the Apollo 13 mission, simple models have been succeeded by computerization of those models, facilitating NASA to screen and alter frameworks in genuine time with ever more precision. But the fundamental idea is still the same even now. It is a demonstration of a physical object termed a 'twin' which empowers one to screen its status, analyze issues and test arrangements remotely. A digital model of a physical device a 'twin' empowers one to screen its status, analyze issues and test functions remotely.

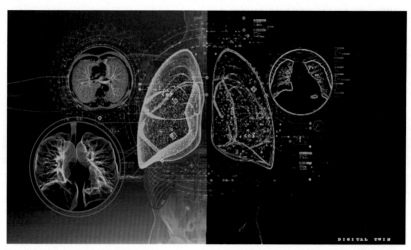

Fig. 3 Digital Twin in healthcare. *Source: https:/www.information-age.com/gartner-digital-twins-123479330/.*

The same concept of replication of the physical model as a digital model has applied the identical conception to healthcare promotion, where healthcare suppliers and patients rest on the precise results of an X-ray device and scanning machines like MRI scanner (Fig. 3).

When the digital representation of the physical object includes within the power of computational advances such as Artificial Intelligence, it will indeed distinguish potential issues immediately they emerge and permitting for well-timed repair or substitution of basic components. For example, a profound investigation of information transmitted from sensors in a flight engine during flying can give 15–30 days' advance note of potential failures. In this scenario, it is evident that the healthcare functionalities can reap maximum advantage from the same kind of prognostics.

3.1 Identify maintenance needs before they arise

In situations like the cancelations or postponement of the medical tests prescribed for the patients and unexpected workflow, instabilities are basic issues for both clinics and patients. Imaging systems ought to be well prepared and operational whenever they require it. Framework disappointments can cause spur-of-the-moment downtime of the equipment and when it is beyond the stipulated waiting time will cause inconvenience, with a potential negative influence on clinical results as a ripple effect. It is also not so easy to give the required support immediately in real time.

For an instance, one ought to replace the seat belt in the car or the chain on a bike after using for a while and it is the same situation where certain components of an MRI scanner degrade over time through standard utilization and demands replacement or repair. The challenge, at this point, is to recognize and predict potential issues well in advance before they occur, so it is easy to plan support at a time when the device is not in use usually most of the time at night.

This is where the concept of the digital twin comes in the field of healthcare. Each day, an ordinary MRI scanner produces a normal of 800,000 log messages, which reflect how the framework is working actually. It is mandatory to have proactive remote observing administrations of these logs to track and analyze these log messages for early caution signs of forthcoming specific issues.

The following are the common questions arise with respect to medical emergencies.

- Whether a patient requires hospitalization?
- What are the alternatives if a patient has a surgical procedure?
- What types of care a patient needs and how many days of hospital stay if the patient needs hospitalization?
- How long it requires to be healed completely and what is the level of positive intervention of the medicines the patient prescribed to?
- What is the follow-up duration of the patient?
- How long and how much in terms of dosage medicine to be given?
- What can be the diet and alteration of the lifestyle?
- What are the side effects of the post-treatment?

These questions can be answered with the assistance of Digital twin as a trustworthy and efficient support system to a medical practitioner. The following simple scenario will help us to understand the need.

Let us consider an example of a diabetic patient whose blood sugar levels to be monitored at regular intervals and alter the medicine according to the levels is the routine strategy. There are advanced digital devices flowing into markets to have continuous measurements of the glucose level in the blood in an easy manner. Still, the patient requires the support of a medical practitioner often to have the medicine altered. There are situations where the patient can avoid the intervention of a doctor with the support of the technology. Let us imagine that there is a technological assistance to a patient with the input of the daily routine alterations, previous drug interventions during the alter of the glucose levels, diet of the patient with glucose level in it and alteration of inherited habits with the input of changing all these according to the current glucose level in the blood. Indeed it will work as an expert medical practitioner. Of course, it can be achieved through Digital Twin platform which utilizes all this information in the context of the patient and can act as an expert system.

Following are some of the benefits which can be reaped out of digital twins in healthcare.

(a) *Nothing less nothing more*

There are possibilities for errors in billing like adding extra cost or missing what to be charged or changing someone else's bill due to the human error. It may happen due to the overcrowding, human exhaustion, system failures or network issues. Inclusive of IoT in healthcare with Digital Twins can fix all these human errors with respect to supplies, services, people and process.

(b) *Optimized waiting time for patients*

A digital twin also plays a big part in optimizing day-to-day performance. Because of this facility, data collection, data storage, and data retrieval can happen error free as well as fast manner which will significantly reduce the waiting time of the patients. When a patient is getting admitted, finding the right sources such as doctors, rooms, availability of the external sources and the required facilities can be streamlined within a very short time.

(c) *Precise inventory*

Because of Digital Twin, the data availability of the real-time consumption and arising need for any new medicines through prediction can keep the store all the time up to date. Meanwhile discarding the medicines which are merely used or unused can also go out of the inventory.

(d) *Better surroundings*

Since hospitals are vulnerable areas where most cleanliness and care is expected, it would be highly appreciated if the administration gets prediction through Digital Twin of any epidemic diseases based on the past, the precautionary measures can be geared up to promote a healthy environment.

3.2 Digital heart twin

Instead of using a scalpel in an operation theater for a heart, Dr. Benjamin Meder, the cardiologist at Heidelberg University Hospital in Germany, used mouse and monitor to place the electrodes of a pacemaker with utmost care in a beating digital heart. It has been made possible with the help of AI coupled Digital Twin. Meder has gone through the simulations of the heart of the particular patient before starting the surgery to verify that the pacemaker can help the sufferer alive who went through congestive heart failure.

The Siemens Healthineers has developed the digital heart twin using Artificial Intelligence is the best example that how digital twins coupled medical devices help doctors to create more accurate diagnoses.

The increasing expenses for the healthcare especially surgeries related to heart which costs huge expenses can be avoided with the AI enabled medical tools with digital twin by finding if the surgery is unnecessary. It can save thousands of dollars, patient's health and time very effectively.

Meder says from his experience that it would be possible with digital twins to foresee the health status of the patients weeks or months before and it would be analyzed and understood the response level of the patient to the particular treatment through which the patients' health is ensured.

4. Smart cities

Nowadays the governments are coming up with the project for smart cities to enable complete facilities to the people without any delay or interruptions. As it is defined earlier, the Digital Twins create the three-dimensional virtual representations of constructions, infrastructure, and other physical assets connected along with the data in and around them. As it is the primary duty of any government to administer efficiently the functional areas like maintenance, energy consumption, space utilization, traffic management, and public safety in any city, Digital Twins can be used. Many governments have come forward to use Digital Twin with Machine Learning and Artificial Intelligence to build smart cities in the world.

A little town within the Swiss Alps has witnessed the emerging of the digital twin of one of the cities with the advancement in technology within the world. The team working on building smart cities for Switzerland is led by Nomoko which is a Zurich start-up and the Swiss Federal Railways, along with the other companies such as Swisscom, Swiss Post, and AMAG. The venture implements an application platform to construct 3D models of a city. This 3D model helps to create and assess the communication and administration of diverse structures of the city such as movement, communications, energy producing and consuming systems, etc. The first model of the smart city using Digital Twin will be fashioned with the idea of the city of Basel.

Amaravati, the modern capital of the Indian state of Andhra Pradesh, is thought to be the primary whole city born with a Digital Twin. The preliminary 3D model of the city was built by Cityzenith using Smart World Pro software, has been revealed at the World Financial Forum's yearly common assembly in Davos, which took place in January 2019. Michael Jansen, CEO of Cityzenith, has described that the whole lot that occurs in Amaravati will be predicted well before to augment outcomes (Fig. 4).

4.1 The benefits of having a twin for smart cities

In this context, a Digital Twin could be a computerized illustration of a physical resource, which collects data by means of sensors, rambles or other IoT and Mechanical IoT instruments and applies improved analytics, Machine Learning (ML) and Artificial Intelligence (AI) to pick up real-time experiences almost the physical asset's execution, operation or profitability. Such innovation looks set to play a progressively vital part within the creation of smart cities around the world and intending to major open well-being, security, and natural issues.

Bringing the virtual and physical universes together in this way can offer support to way better illuminate decision making, weakens the vulnerabilities additionally acts as a citizen engagement tool. Itron Ideas Labs team is connecting with Microsoft Azure to utilize mixed reality to form a virtual illustration of the networks between building supplies, foundation and different sensor sorts in a downtown Los Angeles neighborhood.

It will utilize a Microsoft HoloLens holographic system and the computer program company's digital twin innovation to appear how organizers and city specialists in Los Angeles can virtually install smart sensors, alter

Fig. 4 Digital Twin in smart cities. *Source: https://www.forbes.com/sites/danielnewman/ 2019/01/08/are-privacy-concerns-halting-smart-cities-indefinitely/#2767b269ba6d.*

housetop materials, change activity designs, plant trees and see the outcomes on individual citizens, cars, schools and buildings within the simulated surroundings.

4.2 Build repeatable, versatile experiences

Itron features a portfolio of smart systems, computer programs, administrations, meters and sensors to assist its clients way better oversee power, gas and water assets for the individuals they serve. Through its Thought Labs, the company needs to utilize developing innovations such as augmented reality and machine learning to assist and advise plans that improve the quality of life for its client base.

Itron is among the primary adopters of Microsoft's Digital Twins benefit, which empowers engineers to construct repeatable, adaptable encounters from digital sources and the physical world. Neighborhood advancements will show within the simulated environment, permitting users to create well-versed choices with respect to technologies they may be considering implementing.

4.3 A digital twin for the city of Newcastle

The post-graduate students from Newcastle College, the United Kingdom working with Northumbrian Water, have made a Digital Twin of the city to assist it to respond in a better manner to occurrences and debacles. The virtual representation permits the water company to run computer-generated simulations of occurrences such as burst pipes, heavy rainfall or severe flooding to illustrate in few minutes the bad effects it may have on people's homes and communities over a period of 24 h.

The Digital Twin will not support the city to respond in difficult times to such freak climate occasions alone, but too to assess an unlimited number of potential emergencies expected in the future. In a circumstance like unexpected heavy rainfall and storms, a digital twin of the city would have been exceptionally valuable. It tells us which buildings will be overwhelmed, which foundation will be closed down, clinics that may be influenced. It gives us a picture promptly of which ranges will be affected.

The developers of this project make clear that in the event that an occurrence does happen, it can work with crisis responders to run re-enactments for any area and recognize issues faster and easier with this digital twin smart city.

4.3.1 Introducing Digital Twin citizens

Whereas digital twins have the potential to be massively profitable for organizers and city specialists, expanding the business communities with other partners who can augment the involvement in smart city projects. Amaravati and CityZenith are proposing a digital twin client ID associated with each resident that will assist as a sole point of access for all government data, intimations, facilities, and applications. The usage of digital twin id will offer assistance to make progress in networking with both citizens and the private sector.

Sreedhar Cherukuri, commissioner, Andhra Pradesh Capital Regional Development Authority (APCRDA), has stated that Amaravati could be a Greenfield city built with the safety and security of its residents at the center of its dream.

5. Automobile

The automotive business is shifting swiftly as the time demands along with the novel technologies, increasing and varying consumer needs, and new vendors come into the business space. The industry is no longer work alone with the manufacturing hardware in the traditional way but stepping into the field of coupled mobility and conveyance solutions, stretching from services coupled with emerging digital services, operated with sensors, artificial intelligence-enabled devices, automotive means of transportation of the future. Despite the fact new contributions materialize, the challenges to create the value of the business and increasing the profit has made a demand from executives to associate with the technological fields in which there is a great challenger may take advantage. In the automobile industry digitalization can be viewed as a reagent, architect, and enabler for automotive and mobility businesses. The involvement and importance of digital twin in a hardware enhancement, constructing and value life cycle of the vehicle are incredible (Fig. 5).

5.1 Innovation in automotive product

The involvement and significance of digital twin have made it as an essential framework in developing a product, engineering and also in the service life cycle of the vehicle.

The automotive product innovation has come through a long way and complex from the manual creation. Ordinarily, a modern car model needs 5–6 years of time from a design phase to launch phase. In reality, this period

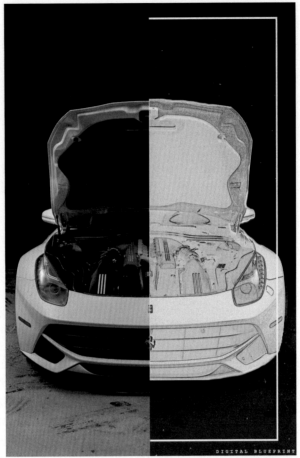

Fig. 5 Digital Twin in automobile. *Source: https://new.siemens.com/id/en/company/topic-areas/digital-plant.html.*

is very crucial since the success of the model and its robustness to sustain in the business market for the long term. If a minor error is overlooked within the product plan can disintegrate the company's net worth in the market and its turnover. For example, one of the original equipment manufacturer (OEM) introduced their new product at the beginning of the year 2000 with the development expense of 1.5 billion dollars. But the model became unsuccessful in a moose test to understand its shock absorbent power because which they had to withdraw almost 2500 new cars from the market. The company added stability control mechanism to the model and reformed the car's suspension with the implementation cost of approximately 250 million dollars to regain its market value.

In every stage of the product development life cycle, the design engineers and product engineering team faces various key issues to be solved. Digital Twin can represent the actual stage and the key activities performed in the stage to address these challenges.

5.2 Engineering of vehicle

Henry Ford's advancement in car manufacturing for more than a century ago reduced the time it consumed 2 h and 30 min to manufacture a car which was originally more than 12 h. Since at that point, the business has seen different turbulences and advancements also. In the current scenario, every 30 s, a new car is coming out of the garage queue because of the advancements. The machine beneath the cover has advanced from an unassertive mechanical wonder to a multifaceted and brilliant framework consist of a bunch of technologies, gadgets, and resources. A quick and refined manufacturing implementation depends on the strength of administration of various resources, development proposal, and procedure monitoring. Models and their variations have amplified manifold and customizations of the vehicles have similarly increased in essence. The wider vision of whole product development life cycle and weight on progressing the OEE features like 'first time through' has taken an important place on digital engineering among all the vehicle producers nowadays.

Well planned and executed digital adoption is now evolving as a critical accomplishment aspect for the manufacturing. This happens by collecting and investigating more data in a virtual framework which will give superior choices and, in many cases, even the prognostic choices can be made. Digital twin brings out the mighty possibilities into the picture and addresses the typical challenges in the manufacturing cycle.

5.3 Vehicle sales and service

The revolution in the field of research, manufacturing products, network development, advertising operations and a massive struggle of over 5 years of time for a new vehicle introduction into the market brings the best product at retail which decodes these investments into income for the Producer. The profits after the sales from individual spare parts, fixtures, and services are also hooked on the sales in the market.

Current automobile sales ground is perceiving numerous drifts and pattern swings with the evolving service-oriented model, purchaser need for higher and customized user experience at marketing and single-channel experience for every single dealing in a transaction, governing acquiescence of the particular territory, etc. Auto Producers functioning at an International scale have an even greater challenge of dealing with macro conservational parameters and geographical individualities. OEMs are powerful to extract the best functional insights from the consumers, vehicle and network associates to unceasingly progress product performance. But due to various inadequacies, restrictions and outside influences, these appreciated perceptions get worn.

In this scenario, the digital representation which comprises the complete life cycle of an automotive product with the support of digital twin will predict the manufacturing issues, market trend changes and the customer demand fluctuations well in advance and alter the way the production happens. The digital twin will help to overcome the issues related to boundaries, networking partners and the territorial compliances and help the product to infiltrate the automobile market efficaciously.

The advancement in analysis, designing artfulness, organizing schedules, advertising operations and a colossal exertion of over 5 years time for a modern vehicle production comes down at a retail industry which deciphers these ventures into income for the producer. The aftersales income from spare parts and accessories are good on genuine sales whereas administrations will cost very little. Modern auto deals floor is seeing different patterns and world view shifts with developing new models such as a client request for prevalent and personalized client involvement at retail and omnichannel involvement for each value-based interaction, administrative compliance like GDPR, etc.

Auto producers working at a worldwide scale, face an indeed greater task of managing with large-scale natural variables and geological quirks. OEMs are sharp to use the operational bits of knowledge from the clients, vehicle and channel accomplices to ceaselessly make strides item execution. But due to different wasteful aspects, imperatives and outside components, these profitable experiences get disintegrated.

6. Retail
6.1 From transactions to engagement

Retailers are transforming their move from transaction-focused intelligent to supported consumer engagement, identifying that consumers' purchasing behavior does not end with the buying alone whether it is a visit to the store or buying online through portals but it is the continuous journey accommodating the customer as a primary person in the business. In the current trend of online or offline shopping associated with various portals like Amazon, Flipkart or Walmart, etc., it is evident that attracting customers is not alone the factor to sustain in the business but also to retain them by providing the best customer support. Customers are equipped with the knowledge of the product and it's availability in different portals to the best price and faster delivery. The search engines like Alexa, Google, and Siri are popping with advertisements of new and similar products which customer may need with the help of big data analytics to grab the customer. But it is clear that there is a gap in meeting the retailers' promotions and the customers' needs and that is where the retailers try their best to reach the customers and make them buyers. Digital Twins can render 360-degree analysis and enhance the solutions to bridge the customer needs and retailers reach (Fig. 6).

6.2 Creating a Digital Twin of your customer

As a customer is linked with various products ranging in different aspects of usage and building a Digital Twin of a customer will definitely help in improving retail management.

Industrial products enterprises have been creating 'digital twins' of their significant assets such as manufacturing equipment, mining devices, transport automobiles, large-scale freezers and industrial ovens that observe and measure functioning performance, temperatures, heat production, and wear and tear. By mining the information from their resources in real-time and including the artificial intelligence methodologies to efficaciously

Fig. 6 Digital Twin in retail. *Source: http://www.whichvoip.co.za/news/gartner/gartner-predicts-top-10-global-retailers-will-use-real-time-store-pricing-2025.*

foresee the issues and proactively address the required support it is easy to avoid the impromptu downtimes. The implementation of digital twins has effectively augmented efficiency and profit, in addition to creating absolutely idle business models.

Retailers and Consumer Packaged Goods (CPG) companies presently have an equivalent opportunity. By combining nitty-gritty value-based information with experiences from social media and utilization of information from IoT smart devices like smartphones, smart home lighting, machinery, domestic entertainment systems, heating, and cooling systems, and smart cars, it is presently getting to be feasible to form 'digital twins' of consumers.

We can consider the example of the behavior of a customer buying a particular brand coffee. Combining value-based information such as the amount and recurrence of coffee purchases with bits of knowledge to consumption, as well as non-transactional information such as family measure and makeup, the time the family is out of home, spending time in the office or in an excursion, etc., will enable the companies to construct a reliable model of digital twin for each consumer, to forecast timing and recurrence of purchase of a particular item. As each customer is working with his or her own time frame of reorder point, where it would be beneficial for the customer to get the advertisement of the particular brand with an offer from the retailer.

Applying these bits of knowledge with machine learning to get the customers' behavior, even it would be easy for the retailer to show the nearest outlets where the customer can get the desired brand product with best possible offers. It will benefit a win–win situation for the customers as well as the retailers.

Cross-sell and up-sell are possible for the retailers by leveraging intuitions received from both the user experience of the products as well as the utilization of the associated device. For an instance, when there is an increase in the need for a low calories food to be prepared in some circumstances like hospitals, with the above insights digital twins can offer completely new prospects for the retailers to cross-sell and up-sell.

In another example of digital twins for a customer, we can consider a person who is leaving a car on for a time with its air-conditioning is on or the headlights on during daytime while it is absolutely not necessary. In such situation, the prescribed inner battery life may come down and with the customer digital twin, it can be resolved by giving a suggestion to the customer or providing an alert before the time of battery drainage to improve the life of the battery.

Considering the current customer-centric approach in any business, it would be highly beneficial to apply 'Digital Twin' concepts to predictively offer significant suggestions to the customers for the products which may be of their priority. It will surely make the customers fully satisfied and attract more often and it will increase the benefit for the retailers also.

6.3 Digital Twins to benefit retail industry

Since in the retail or Internet of Things, the Digital Twins represents the computerized replica of an item on the web that can be altered and refurbished as the genuine physical item goes through diverse stages of its life cycle. As the life cycle of a product starts from assembling, manufacturing, distribution, sales to client consumption, experiences and the expiry of the product, etc., these digital twins keep advancing and upgrading their item status as they move on upward within the generation cycle. This real-time update is having control over the entire life cycle of a product which predicts and solves the issues well in advance.

In retail, data twins may be utilized to get the data refined at all stages of their life cycles. In addition to that, by building IoT associated smart products or digital twins, products can diminish the hazard of deterioration and cut down the turnaround time. This is often the case, particularly with consumer-packaged goods. Products can too upgrade offers or special changes or upgrade other information related to the physical item and its packaging. Unlike a simulation in manufacturing, digital twins in retail are more important to track and follow items, appropriate to logistic firms and supply chain managers. This data-driven digital twin data can be shared among partners, organizations, groups and indeed nations and can too be conveyed in a multilingual procedure. It endows businesses as well as the end users to keep track of items and their associating data as they are sourced, conveyed, obtained, utilized or consumed.

Digital twins pave the way for all the information as discounts, product batch, new products on certain instances, supply chains the products have gone over, place of the product at any point of time and the last packaging time to be stored digitally and accessed over the Internet throughout the product's entire lifecycle. This will assuredly expose the new doors to immense potentials for the retail applications. In conclusion Digital Twins brands customer retail extra competent, dropping marginal expenses and letting extra grainy scrutiny of data to progress customer experiences and enhancing productivity.

Further reading

[1] Tao F, Cheng J, Qi Q, Zhang M, Zhang H, Sui F: Digital twin-driven product design, manufacturing and service with big data, *Int J Adv Manuf Technol* 94:3563–3576, 2018.

[2] Glaessgen E, Stargel D: The digital twin paradigm for future NASA and US Air Force vehicles. In *53rd AIAA/ASME/ASCE/AHS/ASC Structures, Structural Dynamics and Materials Conference*; 2012, p 1818.

[3] Haag S, Anderl R: Digital twin–proof of concept, *Manuf Lett* 15:64–66, 2018].

[4] Mohammadi N, Taylor JE: Smart city digital twins. In *2017 IEEE Symposium Series on Computational Intelligence (SSCI)*; 2017, pp 1–5.

[5] Canedo A: Industrial IoT lifecycle via digital twins. In *Proceedings of the Eleventh IEEE/ACM/IFIP International Conference on Hardware/Software Codesign and System Synthesis*; 2016, p 29.

[6] Tao F, Zhang M, Nee AYC: *Digital Twin Driven Smart Manufacturing,* first ed., 2019, Academic Press.

[7] David Stephenson W: *The Future is Smart: How Your Company Can Capitalize on the Internet of Things—and Win in a Connected Economy,* 2018, AMACOM.

[8] Copley C: Medtech firms get personal with digital twins, *Science News,* 2018. August 31, https://in.reuters.com/article/us-healthcare-medical-technology-ai-insi/medtech-firms-get-personal-with-digital-twins-idINKCN1LG0S0.

[9] Watts B: Digital Twins Virtualizing Hospitals, 2018: https://www.challenge.org/knowledgeitems/digital-twins-virtualizing-hospitals/.

[10] https://gizmodo.com/this-is-the-actual-hack-that-saved-the-astronauts-of-th-1598385593n.

[11] *Digital Twins in Healthcare: Enabling Mass Personalization of Care Delivery,* 2019, Persistent Systems Ltd https://www.persistent.com/wp-content/uploads/2019/02/digital-twins-whitepaper.pdf.

[12] Luterbacher C: Why Our Cities Aren't as Smart as They Could Be, 2018: https://www.swissinfo.ch/eng/energy-innovation_why-our-cities-aren-t-as-smart-as-they-could-be/44359108.

[13] Cityzenith: SmartWorldPro, 2019: https://republic.co/cityzenith.

[14] Glocker G: *A Primer on Digital Twins in the IoT,* 2018, Bosch IoT Suite https://blog.bosch-si.com/bosch-iot-suite/a-primer-on-digital-twins-in-the-iot/.

[15] McCarthy D: 5 Reasons Digital Twins Matter to Your IoT Deployment, 2018: https://www.networkworld.com/article/3253891/5-reasons-digital-twins-matter-to-your-iot-deployment.html.

[16] GE Power Digital Solutions: *GE Digital Twin: Analytic Engine for the Digital Power Plant,* 2016, GE Power Digital Solutions https://www.ge.com/digital/sites/default/files/download_assets/Digital-Twin-for-the-digital-power-plant-.pdf?gecid=extref_dig_gepowercom.

[17] Millman R: Gartner: Four Best Practices for Managing Digital Twins, 2018: https://internetofbusiness.com/half-of-businesses-with-iot-projects-planning-to-use-digital-twin/.

[18] Quirk E: Gartner Survey Reveals Increased IoT Implementations and Use of Digital Twins in 2018: https://solutionsreview.com/business-process-management/gartner-survey-reveals-increased-iot-implementations-and-use-of-digital-twins-in-2018/.

[19] Siemens: Digital Enterprise for process industries—Start your digital transformation now, 2019: https://new.siemens.com/global/en/company/topic-areas/digital-enterprise/process-industry.html.

[20] Ismail N: Gartner: Digital Twins Beginning to Enter the Mainstream, 2019: https://www.information-age.com/gartner-digital-twins-123479330/.

[21] Newman D: Are Privacy Concerns Halting Smart Cities Indefinitely?, 2019: https://www.forbes.com/sites/danielnewman/2019/01/08/are-privacy-concerns-halting-smart-cities-indefinitely/#2767b269ba6d.
[22] Egham, Gartner Predicts Top 10 Global Retailers Will Use Real-Time In-Store Pricing by 2025, 2019. Retail: http://www.whichvoip.co.za/news/gartner/gartner-predicts-top-10-global-retailers-will-use-real-time-store-pricing-2025.

About the author

 **Peter Augustine** is currently serving as an Associate Professor at Department of Computer Science, Christ University since 2005. Previous to this he served as the Research Executive at Techsoft Private Ltd., Secunderabad for 2 years. Dr. Peter received his BSc and MCA degrees from St. Xavier's College, which is currently accredited as A++ by NACC and rated as number 1 in India. He completed his MPhil in Computer Science from Manonmaniam Sundaranar University, Tamilnadu and completed his PhD from CHRIST (Deemed to be University). Dr. Peter has published over 20 research papers in international journals. He has been doing Major Research Project on Brain Tumor Detection using CNN sponsored by CHRIST University with the fund of above 18 lakhs. He has also been working for a research project along with super speciality doctors of pulmonology at St. John's Medical Research Institute, Bangalore for the identification of lung disease. He has been working with Big Data Analytics using Hadoop for the past 4 years and deployment of the same applications in cloud environment.

CHAPTER FIVE

Digital twin: Empowering edge devices to be intelligent[☆]

Vidya Hungud, Senthil Kumar Arunachalam
Reliance Jio Infocomm Ltd, Mumbai, India

Contents

[☆]Assuming audience are tech savvy and is a look ahead reader, in this chapter we will walk you through a case study that gives more clarity into the space of enabling digital twin at the edge.

Advances in Computers, Volume 117
ISSN 0065-2458
https://doi.org/10.1016/bs.adcom.2019.10.005

Abstract

Edge/fog devices, such as machineries at manufacturing floors, instruments in hospitals, digital assistants with human beings, equipment and appliances at factories, wares and utensils at homes, cameras at important places such as airports, railway stations, retail stores and malls, entertainment plazas, eating joints, stadiums, and auditoriums are all set to join in the mainstream computing as they are being stuffed with increased processing capabilities, memory and storage capacities. They are also being connected with one another in the vicinity directly and indirectly through a middleware solution. Also remotely held, cloud-hosted and cyber applications and databases are also being integrated with ground-level edge/fog devices. Through such extreme connectivity and deeper integration, edge and fog devices are intrinsically and externally enabled to do real-time data capture, processing, analytics, knowledge discovery, decision-making and actuation. Precisely speaking, edge/fog devices and their clusters are all set to become the next-generation, highly optimized and organized IT infrastructure for producing and delivering real-time applications and services. In other words, edge/fog clouds are being dynamically established and sustained in order to provide a variety of service-oriented, event-driven, people-centric, mission-critical, knowledge-filled, and context-aware services for both professionals and commoners.

Real-time sensor and streaming data analytics can be easily accomplished through edge clouds in order to supply personalized, predictive and prescriptive insights. Edge clouds are also getting synchronized with conventional clouds such as public, private, and hybrid clouds in order to facilitate comprehensive and historical data analytics. This chapter is for accentuating and articulating how edge devices become intelligent in their actions and reactions in conjunction with their respective digital twins.

1. Introduction

The device ecosystem consistently is on the growth path. We are being bombarded with scores of slim, sleek and smart, handy and trendy, purpose-specific and agnostic, resource-constrained and intensive, personal and professional devices. It is forecast by the leading market researchers and analysts that there will be billions of connected devices in the years to unfold. There is no doubt that a number of strategically sound innovations and disruptions are happening in the device space. Distinctly, there is a faster proliferation of multifaceted devices, which are elegantly wearable, implantable, pocketable, and portable. Further on, devices are meticulously empowered to be computational, communicative, sensitive, responsive, perceptive, decision-taking, and active.

With the edge/fog computing continuously gains prominence for ensuring real-time data capture, processing, analytics, decision-making, and actuation, the future cloud is primarily made out as a dynamic and ad hoc collection of edge devices. That is, forming and sustaining edge device clouds is beset with a number of technical challenges and concerns. With the billions of connected devices, the world is leaning toward the realization of pioneering and powerful device clouds to tackle specialized business requirements and to solve the problems, which are not being solved through the current technologies. Also, to bring forth a bevy of real-world disruptions and transformation, edge device clouds are being touted as the way forward. Precisely speaking, the real and sustainable digital transformation can be realized and rewarded through edge/fog device clouds formation and sustenance.

In IoT applications, we will want to track behavior of each individual edge device. The software that helps in abstracting this out, combining event handling and state information about edge device, and enabling introspection on dynamic status of device will enable a win-win for digital twin adoption through digital transformation.

1.1 The evolution of edge computing

The faster maturity and stability of edge technologies has blossomed into a big factor in realizing scores of digitized elements/smart objects/sentient materials out of common, cheap and casual items in our midst. These empowered entities are data-generating and capturing, buffering, transmitting, etc. That is, tangible things are peppered with and prepared for the future. These are mostly resource-constrained and this phenomenon is called the Internet of Things (IoT). Further on, a wider variety of gadgets and gizmos in our working, walking and wandering locations are futuristically instrumented to be spontaneously interconnected and exceptionally intelligent in their behaviors. Thus, we hear, read and even feel connected and cognitive devices and machines in our everyday life. Once upon of a time, all our personal computers were connected via networks (LAN and WAN) and nowadays our personal and professional devices (fixed, portables, mobiles, wearables, implantable, handhelds, phablets, etc.) are increasingly interconnected (BAN, PAN, CAN, LAN. MAN, and WAN) to exhibit a kind of intelligent behavior. This extreme connectivity and service-enablement of our everyday devices go to the level of getting seamlessly integrated with off-premise, online, and on-demand cloud-based applications, services, data sources, and content. This cloud-enablement is capable of making ordinary devices into extraordinary ones. However, most of the well-known and widely used embedded devices individually do not have sufficient computation power, battery, storage and I/O bandwidth to host and manage IoT applications and services. Hence performing data analytics on individual devices is a bit difficult.

As we all know, smart sensors and actuators are being randomly deployed in any significant environments such as homes, hospitals, and hotels in order to minutely monitor, precisely measure, and insightfully manage the various parameters of the environments. Further on, powerful sensors are embedded and etched on different physical, mechanical, electrical and electronics systems in our everyday environments in order to empower them to join in the mainstream computing. Thus, not only environments but also all tangible things in those environments are also smartly sensor-enabled with a tactic as well as the strategic goal of making them distinctly sensitive and responsive in their operations, offerings, and outputs. Sensors are sweetly turning out to be the inseparable eyes and ears of any important thing in-near future. This systematic sensor-enablement of ordinary things not only make them extraordinary but also lay out a stimulating and sparkling foundation for generating a lot of usable and time-critical data. Typically sensors and sensors-attached assets capture or generate and transmit all kinds of data to the faraway cloud environments (public, private, and hybrid) through a host of standards-compliant sensor gateway devices. Precisely speaking, clouds represent the dynamic combination of several powerful server machines, storage appliances, and network solutions and are capable of processing tremendous amounts of multistructured data to spit out actionable insights.

However, there is another side to this remote integration and data processing. For certain requirements, the local or proximate processing of data is mandated. That is, instead of capturing sensor and device data and transmitting them to the faraway cloud environments is not going to be beneficial for time-critical applications. Thereby the concept of edge or fog computing has emerged and is evolving fast these days with the concerted efforts of academic as well as corporate people. The reasonably powerful devices such as smartphones, sensor and IoT gateways, consumer electronics, set-top boxes, smart TVs, Web-enabled refrigerators, and Wi-Fi routers are classified as fog or edge devices to form edge or fog clouds to do the much-needed local processing quickly and easily to arrive and articulate any hidden knowledge. Thus, fog or edge computing is termed and tuned as the serious subject of study and research for producing people-centric and real-time applications and services.

1.2 Briefing of fog/edge computing

Traditional networks, which feed data from devices or transactions to a central storage hub (data warehouses and data marts) can't keep up with the data volume and velocity created by IoT devices. Nor can the data warehouse model meet the low latency response times that users demand. The Hadoop platform in the cloud was supposed to be an answer. But sending the data to the cloud for analysis also poses a risk of data bottlenecks as well as security concerns. New business models, however, need data analytics in a minute or less. The problem of data congestion will only get worse as IoT applications and devices continue to proliferate.

There are certain interesting use-cases such as rich connectivity and interactions among vehicles (V2V) and infrastructure (V2I). This emerging domain of IoT requires services like entertainment, education, and information, public safety, real-time traffic analysis and information, support for high mobility, context-awareness and so forth. Such things see the light only if the infotainment systems within vehicles have to identify and interact with one another dynamically and also with wireless communication infrastructures made available on the road, with remote traffic servers and FM stations, etc. The infotainment system is emerging as the highly synchronized gateway for vehicles on the road. Local devices need to interact themselves to collect data from vehicles and roads/expressways/tunnels/bridges to process them instantaneously to spit out useful intelligence. This is the salivating and sparkling foundation for fog/edge computing.

The value of the data decreases as the time goes. That is, the timeliness and the trustworthiness of data are very important for extracting actionable insights. The moment the data gets generated and captured, it has to be subjected to processing. That is, it is all about real-time capture. Also, it is all about gaining real-time insights through rule/policy-based data filtering, enrichment, pattern searching, aggregation, knowledge discovery, etc. to take a real-time decision and to build real-time applications. The picture below clearly articulates how the delay in capturing and analyzing data costs a lot in terms of business, technical and user values.

The latest trend of computing paradigm is to push the storage, networking, and computation to edge/fog devices for availing certain critical services. As devices are interconnected and integrated with the Internet, their computational capabilities and competencies are uniquely being leveraged in order to lessen the increasing load on cloud infrastructures. Edge devices are adequately instrumented at the design stage itself to interconnect with nearby devices automatically so that multiple devices dynamically can be found, bound, and composed for creating powerful and special-purpose edge clouds. Thus, the concept of fog or edge computing is blooming and booming these days.

The essence and gist of fog computing are to keep data and computation close to end-users at the edge of the network and this arrangement has the added tendency of producing a new class of applications and services to end-users with low latency, high bandwidth, and context-awareness. Fog is invariably closer to humans rather than clouds and hence the name "fog computing" is overwhelmingly accepted across. As indicated and illustrated above, fog devices are typically resource-intensive edge devices. Fog computing is usually touted as a supplement and complement to the popular cloud computing. Students, scholars, and scientists are keen toward unearthing a number of convincing and sellable business and technical cases for fog computing. Being closer to people, the revitalized fog or edge computing is to be extremely fruitful and fabulous in conceptualizing and concretizing a litany of people-centric software applications. Finally, in the era of big, fast, streaming and IoT data, fog/edge computing can facilitate edge analytics. Edge devices can filter out redundant, repetitive and routine data to reduce the precious network bandwidth and the data loads on clouds. Fig. 1 vividly illustrates the fast-emerging three-tier architecture for futuristic computing.

The digitized objects (sensors, beacons, etc.) at the lowest level are generating and capturing poly-structured data in big quantities. The fog devices (gateways, controllers, etc.) at the second level are reasonably blessed with computational, communication and storage power in order to mix, mingle and merge with other fog devices in the environment to ingest and accomplish the local or proximate data processing to emit viable and value-adding insights in time. The third and final level is the faraway cloud centers. This introduction of fog devices in between clouds and digitized elements is the new twist brought in toward the ensuing era of knowledge-filled services. Fog devices act as intelligent intermediaries between cloud-based cyber/virtual applications and sensor/actuator data

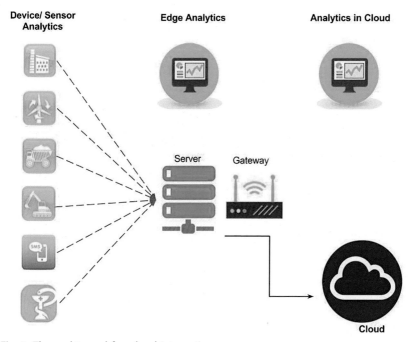

Fig. 1 The end-to-end fog cloud integration.

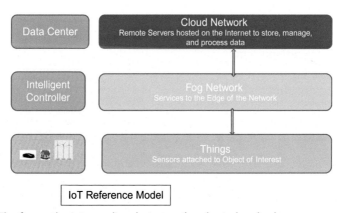

Fig. 2 The fog as the intermediary between the physical and cyber.

at the ground-level. Here is another representation of fog computing as articulated in Fig. 2.

Digital transformation is about marrying IT and business changes, leveraging digital technologies to ensure customer delight by embracing cutting edge technologies that are flexible in their architectural pattern, and have a keen insight in to business process optimization. Digital data is the new fuel for all the business evolutions.

2. Transitions in IoT space

o   Digitization, miniaturization, and consolidation leads to the digitization of physical, mechanical, and electrical systems—device ecosystem is growing rapidly. We are bombarded with a variety of purpose-specific as well as agnostic wearable, handhelds, portable, nomadic, implantable, wireless and mobile devices.

o   Devices are disappearing—we have miniaturized yet multifaceted, invisible yet important, disposable yet indispensable devices for our daily living and working through SoC, micro and nano electronics technologies.

o   Deeper connectivity and extreme integration—sensors networking, device-to-device (D2D) and device-to-cloud (D2C) integration. Thereby we hear the Internet of Things (IoT), cyber physical systems (CPS), device clusters and clouds, etc. often.

o   Gateways/brokers as fog devices—these are single board computers (SBCs) with communication modules, multisensor data fusion capability, and can run application-enablement and analytics platforms.

o   IoT data analytics platforms and applications are being hosted in software-defined cloud environments.

o   The convergence of blockchain, IoT, and artificial intelligence (AI) is coming together for creating secure and sophisticated IoT applications.

3. Data analytics at realtime and @Edge—WHY?

Volume and velocity—ingesting, processing and storing such huge amounts of data which is gathered in real-time.

Security—devices can be located in sensitive environments, control vital systems or send private data. With the number of devices and the fact they are not humans who can simply type a password, new paradigms and strict authentication and access control must be implemented.

Bandwidth—if devices constantly send the sensor and video data, it will hog the internet and cost a fortune. Therefore edge analytics approaches must be deployed to achieve scale and lower response time.

Real-time—data capture, storage, processing, analytics, knowledge discovery, decision-making and actuation.

Less latency and faster response.

Context-awareness—capability.

Combining real-time data with historical state—there are analytics solutions which handle batch quite well and some tools that can process streams without historical context. It is quite challenging to analyze streams and combine them with historical data in real-time.

Power consumption—cloud computing is energy-hungry. It is a concern for a low-carbon economy.

Data obesity—in a traditional cloud approach, huge amount of untreated data are pumped blindly into the cloud that it is supposed to have magical algorithms written by data scientists. This vision is really not the most efficient and it is much more wise to pretreat data at a local level and to limit the cloud processes at the strict minimum.

4. Current trend in edge computing

The emergence of 5G networking and communication capability is to decisively impact on IoT edge analytics and actuation in bringing forth next-generation people and process-centric applications.

The faster maturity of network function virtualization (NFV) and software-defined networking (SDN) are to enable management, utilization and optimization of edge networking resources.

Microservices architecture (MSA) is to realize scores of fog/edge device microservices. The power of machine and deep learning algorithms along with computer vision, natural language processing (NLP) will be made visible in edge device clouds. The overwhelming adoption and adaption of Docker-enabled containerization is to facilitate the deployment of containerized software into edge devices and their networks. Multicontainer edge applications will be the toast of edge computing.

Kubernetes is to manage and orchestrate containerized edge services. Istio and other resiliency frameworks are to help in realizing resilient edge services toward reliable edge environments. The realization of enhanced clouds (the hybrid version of edge and enterprise clouds) is obligatory. The convergence of the blockchain technology and the IoT era promises the IoT security in trust-less environments.

5. Edge computing challenges

- Any IoT environment is hugely dynamic and stuffed with a large number of edge and fog devices. Every device is to be blessed with one or more RESTful APIs for exposing their unique services to the outside world
- Fog/edge device discovery, governance, management, integration, orchestration and security
- Optimal device resource allocation and utilization
- Mapping services/applications with edge device(s)
- Leveraging fog computing for scalable IoT data centers using spine-leaf network topology
- Edge device traffic management, data and protocol translation, etc.
- Forming clouds out of edge and fog devices

6. Why edge cloud analytics?

The traditional cloud centers are doing well in stocking and analyzing big data being emitted by IoT devices, sensors, actuators, etc. and they are good at batch processing and big data analytics. But the brewing requirements is the real-time data capture, storage, processing, analytics, knowledge discovery and actuation. That is, the next-generation real-time analytics and applications cannot be facilitated by the conventional cloud environments.

For guaranteeing real-time analytics and action, we need specialized localized and ad hoc clouds that ensure proximity processing and real-time data analytics. With IoT devices are fast proliferating across the various industry verticals, the future yearns for real-time processing of data being emitted by various types of IoT devices. The current shift is tuned toward producing edge/fog clouds out of edge devices. (As per the market research reports, there will be 50 billion connected devices.) (Fig. 3).

6.1 Advantages

- Real-time processing of data, with proximity to devices, reduces latency in taking actions.
- Reduce the volume of data sent over the wire to public cloud results in the following gains:
  - o Reduced network bandwidth requirement for sending data, which is a costly affair (nearly 80% of the data can be reduced).

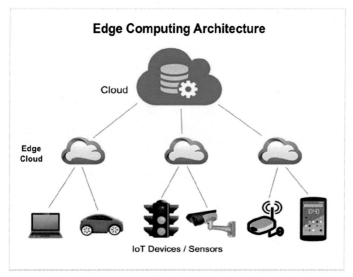

Fig. 3 Edge computing.

o Reduced storage needs for processing data on the public cloud. Storing/retrieving data is very expensive.

o Reduced compute capacity needs on the cloud side for data processing.

o Eliminates the round-trip over the network to make quick/key decisions.

o Localized processing across many edge devices will allow a greater horizontal scalability and reduce cost.

o Containers can help in providing homogeneous, easy to maintain and contained environment for the edge cloud.

7. Edge cloud and its capabilities

Considering the necessity of real-time analytics of IoT edge data, we have developed a sample edge computing environment using Raspberry PI modules to demonstrate the Edge cloud capabilities and its significance. Apache Edgent is the edge analytics platform to be run on Raspberry Pi. This captures sensor data, crunches it immediately and conveys the knowledge to the concerned devices to act intelligently in time. The edge environment is integrated with AWS public cloud for transmitting and processing IoT data leisurely (Fig. 4).

The Macro-level Architecture of Edge Analytics

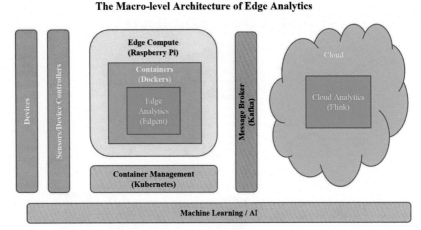

Fig. 4 The macro level architecture of edge analytics.

7.1 Highlights

- How sensors and digitized elements get locally connected with one or more IoT gateway instances in order to gather and transmit any useful and usable data to the IoT gateway. In other words, multistructured and high-volume data getting generated by various sensors and sensors-attached assets in a particular environment (say, homes, hotels, hospitals, etc.) are received and temporarily stocked by IoT gateways/middleware/brokers for purpose-specific data analytics.
- By deploying an edge analytics and application development platform in the IoT gateway (Raspberry Pi was used for our demo), all kinds of data getting collected are getting cleansed and crunched in real-time in order to emit out actionable and timely insights.
- The IoT gateway also contributes in filtering out irrelevant data at the source itself so that a very limited amount of useful data gets transmitted to the faraway clouds to facilitate historical and comprehensive big data analytics. The IoT gateway acts as an intermediary between scores of on-premise edge systems and off-premise clouds.
- IoT gateway modules (typically touted as fog devices) act as the master node/leader in monitoring, measuring and managing various dynamic edge devices and their operational parameters.
- IoT gateway modules seamlessly and spontaneously integrate the physical world with the cyber world (cloud services, applications, databases, platforms, etc.).
- IoT gateway activates, augments, and adapts actuation devices (edge) based on the insights extricated through analytics in real-time.

7.2 Technical stack

- Apache Edgent, Eclipse Kura for edge processing
- Apache flink, spark for cloud processing
- Docker containers and Kubernetes for container orchestration
- Kafka/files for message broker
- Istio—service mesh for resiliency
- IoT devices/Raspberry Pi (Pi4j, wiringpi)

8. Case study—Edge analytics platform and its capabilities

8.1 Problem statement

Today, there are a variety of applications in industrial space and home automation/security space where the data is captured from the devices/sensors and they are faithfully transmitted to a faraway cloud environment and the captured information gets processed and the actions are sent back to the device controller for execution. If the cloud network is not available, then the real problem emanates. Also, the network latency is a barrier.

8.2 Our objective

To create a new edge platform, where collecting and processing the procured data will happen locally at the edge that is unique and extensible to connect with other systems and sensors in the neighborhood via wireless and wireline communication (Fig. 5).

8.3 How are we achieving this?

A variety of sensors and devices are getting integrated with this Edge controller platform. These controllers can form an edge cloud so that if controller goes down, another controller processes the captured information and does the job faithfully. It has a local in-memory storage. The platform is generic and scalable such that it could be used for variety of industrial use-cases. We could demonstrate the capabilities for an industrial security access system for controlling readers, sensors, alarms and access points. The edge application has been designed and developed as a pool of publicly discoverable, interoperable, and composed microservices. Container is the runtime and K8s is the management platform. Service mesh (Istio) is being planned in order to enhance service resiliency. The Edge controller has a connectivity with the backend cloud database so that pull and push communications are enabled. There are

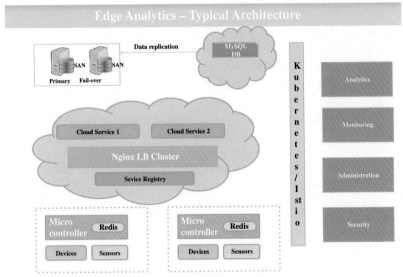

Fig. 5 Edge analytics—typical.

several other features and functionalities embedded inside the controller through microservices, policy files, configuration files, etc.

8.4 USP's

- This is an edge computing-based system. Transitions the data processing and decision-making to the edge device.
- Real-time data capture, storage, processing, and analytics can be done through this edge device so that real-time knowledge discovery and dissemination can be accomplished.
- Advanced deep learning (DL) algorithm has been implemented for making quick and correct analytical decisions on edge cloud.
- The back-end integration with remote cloud systems can be dynamically configured. This device is capable of attending any kind of business events. Multiple Edge controllers can be linked up together to collect and crunch data in an efficient manner.
- Also, machine learning (ML) algorithms are being used at the edge side in order to enable preventive and predictive maintenance.
- Device/User analytics can be performed using streaming analytics capability made available in the edge device. Thus, forming edge clouds out of Edge controllers, performing real-time edge analytics,

assisting for cloud-based comprehensive analytics, the extensibility of the edge device to absorb any kind of futuristic improvements and improvisations, the slim and sleek nature of the controller, the in-memory computing facility, etc. are some of the innovations and disruptions.

- This is a kind of hybrid system. That is, it has an integration with our own private cloud. We can offload some of the complicated workloads to faraway and powerful cloud environments.

Establishing a seamless and spontaneous linkage between edge, private and public cloud environments is greatly simplified through a variety of innovations and industry disruptions. Forming an edge cloud for doing edge analytics for special scenarios is also considered in this project. Machine learning at edge is another game-changer. The controller design is very modular so that any kind of business, technical and user changes and expectations can be quickly and easily met. Thus, the architecture, infrastructure optimization, technology assimilation, process excellence, leveraging of data, etc. are individually and collectively termed as the crucial differentiators and this project has fully met all the initially expressed and envisaged functionalities and advantages. This is one of the highly discussed digital life applications.

9. Cook book/user manual on edge analytics platform

This is to quickly build an edge analytics platform for demonstrating the capabilities of such a platform. We have picked Apache Edgent containerized on Raspberry Pi to perform edge analytics and Apache Flink containerized on AWS cloud to perform server side analytics. Pulse sensor connected to Raspberry Pi can generate pulse data continuously and the sample Edgent App will analyze the data and filter out normal data and pick high outliers for localized processing. Lower/less-significant outliers are then sent to cloud for non-real time analytics on the Flink App.

Both Edgent and Flink have the capability to use a variety of data sources and sink. This demonstration uses files as data sources and sinks. The real platform built for enterprises would use a message broker such as kafka or a simple message queue as an interface between different components. The processed data can be stored in HDFS or DB for summary data.

9.1 Prerequisites
- A cloud machine or a local laptop that is capable of running containers and public internet connectivity
- Raspberry Pi 3+
- Pulse sensor/heart rate sensor
- Analog to digital converter (MCP3008)

9.2 Need to build
- Capture sensor/device data (i.e., pi4j/raspberry pi) or simulate the same
- Edgent
- Flink

9.3 Raspberry PI setup
- Install latest Noobs and Raspbian OS
  - https://www.raspberrypi.org/downloads/noobs/
- Local access requires console (via HDMI) and keyboard/mouse (via USB)
- Configure LAN or WLAN (with this pi can be remotely accessed via SSH)
- Once Booted, update/upgrade packages
- Install/configure pi4j for programming pi to read/write device data

Note
- Requires SD card to be formatted with fat32 prior to download & install of Noobs/Raspbian OS
- General configuration steps for reference:
  - https://www.raspberrypi.org/documentation/configuration/

9.4 Docker
Containerize pi4j, edgent and flink modules for ease of build, management and use.

Note
- Docker should be preinstalled https://docs.docker.com/install/
- For edgent/flink—use image openjdk/alpine (this is light weight)
- Pi4j/Wiringpi requires Raspbian OS

9.5 Pi4j
http://pi4j.com/download.html used to read/write sensor data.

Dockerfile contents for Pi4j

```
FROM resin/rpi-raspbian
    RUN apt-get update \
        && apt-get upgrade \
        && apt-get install wget \
        && apt-get install wiringpi \
        && apt-get install -y oracle-java8-jdk \
        && wget http://get.pi4j.com/download/pi4j-1.2-SNAPSHOT.
        deb \
        && dpkg -i pi4j-1.2-SNAPSHOT.deb \
        && apt-get install pi4j \
        && mkdir /app
    WORKDIR /app
    CMD bash
```

9.6 Sample program to read pulse sensor data

Note

- Containers require Raspbian OS
- Dependency on Wiring pi libraries
- http://wiringpi.com/download-and-install/
- Refer to pi4j documentation for sample programs to read sensor data
- Example code for reading pulse sensor data using python is provided in the below website:
  - https://tutorials-raspberrypi.com/raspberry-pi-heartbeat-pulse-measuring/

9.7 Edgent

Setup Apache Edgent for streaming analytics on the Edge device.

- mkdir -p $HOME/edge-demo/Edgent
- cd $HOME/edge-demo/edgent
- Create Dockerfile under to build a container image
  - *Dockerfile contents below:*
    ```
    FROM openjdk:alpine
        RUN apk update &&\
        apk upgrade && \
        apk add git && \
        apk add bash && \
        mkdir /app
    ```

WORKDIR /app
CMD sh
- Docker build -t demo:edgent.
- docker run –name edgent -d -v /tmp/data:/app/data -it demo:edgent
- cd /app

9.8 Install edgent on the container
- wget   http://mirror.dsrg.utoronto.ca/apache/incubator/edgent/1.2.0-incubating/apache-edgent-1.2.0-incubating-source-release.tar.gz
- tar -xvzf apache-edgent-1.2.0-incubating-source-release.tar.gz
- cd apache-edgent-1.2.0-incubating
- ./mvnw clean install >build.log 2>&1 &
  - o cd /app
  - o git clone https://github.com/apache/incubator-edgent-samples
  - o cd incubator-edgent-samples
  - o git checkout develop
  - o ls
- ./mvnw clean package >build.log 2>&1 &

9.9 Sample program—Pulse sensor aggregation
- cd /app
- download https://drive.google.com/open?id=1w6vXt8cmG2DUSPSe Cp78bTaSaAc5-ISktoedgent-app.tar
- tar -xvf edgent-app.tar

9.10 Apache flink
Setup Apache Flink on the cloud machine for streaming analytics on server side.
- mkdir -p $HOME/edge-demo/flink
- cd $HOME/edge-demo/flink
- Create Dockerfile under to build a container image
- Create Dockerfile under to build a container image
- Dockerfile contents below:
  - o FROM openjdk:alpine
    - ▪ RUN apk update \
      && apk upgrade \
      && apk add git \
      && apk add bash \

&& mkdir /app
- o WORKDIR /app
- o CMD sh
- Docker build -t demo:flink.
- docker run —name flink -d -v /tmp/data:/app/data -p 8081:8081-it demo:flink
- cd /app

9.11 How to install/configure

- wget http://apache.mirror.iweb.ca/flink/flink-1.4.2/flink-1.4.2-bin-scala_2.11.tgz
- tar -xvzf flink-1.4.2-bin-scala_2.11.tgz
- cd /app/flink-1.4.2/bin
- ./start-local.sh

9.12 How to source/process/sink data—Sample program— Pulse sensor aggregation

- cd /app
- download https://drive.google.com/open?id=1w6vXt8cmG2DUSPSe Cp78bTaSaAc5-ISkto flink-app.tar
- tar -xvf flink-app.tar
- cd flink-1.4.2/bin
- ./flink run/app/dist/cloud-analytics.jar
- (Optional) —input=<input file>--output=<output file>
- http://<IP Address>:8081/
  - o Submit new job
  - o —input=<input file>--output=<output file>
  - o Submit job

10. Conclusion

The key to a successful long-term product in market from a technology standpoint, primarily depends on how individual will leverage the futuristic and flexible technologies, selection of right core architectural pattern like MSA (microservices architecture), container orchestration platform like Kubernetes, IMDBG and many more upcoming technologies.

As IoT and IIoT are becoming more and more pervasive, the edge computing will gain traction directly proportional to that. Digital twin will be strategic to enterprise business, as organizations are keen to save data management costs, have tighter control on privacy and security, cut down load on Internet and reduce latency. The combination of edge, cloud and digital twin works really well in IoT and IIoT space. Edge is not a threat to cloud. Industries are willing to scale through device virtualization through virtual twin concepts and also entering into predictive twin through analytical models to reduce their time to market. Organizations have to scope out some time and effort to embrace digital twin model for designing next-generation IIoT and IoT applications.

About the authors

 **Vidya Hungud** is a passionate technologist, product leader, and a keynote speaker with extensive experience (17+ years) in building highly scalable, resilient E-commerce, enterprise applications that are SaaS, IaaS, and PaaS based, including Edge/IoT and cloud infrastructure. After having graduated with master's degree from San Jose State University (SJSU, CA) with major in Client Server Computing, Vidya worked in United States for large size tech company Sun Microsystems on Identity Access Management, mid-size product-based company Intuit on flagship products—TurboTax, QuickBooks, and startup company Software Tree on object-relational mapping software. Vidya has worked on AgriNova, an SMS-based solution to help farmers sell their product at a fair market price that gave an immense satisfaction of having put principles of Design Thinking into practice from ground up. As an innovation catalyst, Vidya has coached several startups at NSRCEL, IIM Bangalore in partnership with Pensaar Inc. and at GHC on Design for Delight and Customer Driven Innovation. As a leader of Tech Women community at Intuit, Vidya, lead 150+ women engineers, was on tech panelist at Cisco, Ericsson. Vidya has contributed as Agile Advisory Board member, has won best manager award at Intuit, and has conducted hands-on workshop on edge computing at OpenStack summit, Vancouver in 2018. Currently, Vidya is driving part of digital transformation journey of India through Reliance, leading Jio Cloud platform and Site Reliability Engineering that entails design through deployment, post-production. As part of paying forward and giving back to community, Vidya has visited

universities and educated on AI/ML, mentored women techies through https://HerSecondInnings.com, speaker in BOAST19 conference, open stack summit, GHCI, IEEE WieILS, WiECon and panelist at Cisco, Ericsson, Pensaar Inc. Vidya continues to drive charter for Tech Women at Reliance Jio (https://www.linkedin.com/in/vidyahungud/).

Senthil Kumar Arunachalam is a seasoned and passionate leader/engineer with around 19 years of industry experience in building and delivering high performance systems, applications, and services on cloud and edge platforms. His key areas of expertise include Edge/IoT, Cloud, Middleware, Databases, and Appliances. Senthil has managed large global teams to drive innovative products, services, platforms, and applications. He is currently serving as Asst. Vice President, owing Performance Engineering and Site Reliability Engineering charter at Reliance Jio Infocomm Ltd., India. Prior to joining Reliance Jio, he has served as a Director of Engineering at Oracle Corporation for over 16 years where he supported Cloud Services delivery with emphasis on performance, scalability, reliability, availability, resiliency, and serviceability. He has earned his engineering graduation in Computer Science and Engineering from Amrita University and did his Executive MBA from Indian Institute of Management, Bangalore. He is an avid analytical thinker and problem solver who applies his engineering, analytical, and financial and business management skills to achieve organizational excellence. He has experience in building teams from scratch, mentored/coached engineers, and managers to deliver optimal results to organizations. He has extensively worked on Edge Analytics, Cloud Infrastructure, Fusion Middleware (Hotspot/JRockit JVMs, Apache/Weblogic HTTP/J2EE Servers, Xen Virtualization stack, Infiniband RDMA Network fabric, ZFSSA Storage arrays and latest Intel IvyBridge/Haswell processors, distributed coherence cache), Business Intelligence, Data Integration (Golden Gate, ODI/ELT, DQ), Monitoring and Diagnostic tools (Enterprise manager, RUEI, Graphite, Logstash, ElasticSearch, Kibana), and Fusion Applications. Besides technology, he enjoys doing yoga, meditation, walking/running, playing cricket, and solving puzzles (https://www.linkedin.com/in/senthil-arunachalam/).

CHAPTER SIX

Industry 4.0: Industrial Internet of Things (IIOT)

Sathyan Munirathinam*

Micron Technology, Manassas, VA, United States
*Corresponding author: e-mail address: sathyan.munirathinam@gmail.com

Contents

Abstract

The physical world is transformed into being digitized and makes everything connected. An explosion of smart devices and technologies has allowed mankind to be in constant communication anywhere and anytime. IoT trend has created a sub-segment of the IoT market known as the industrial Internet of Things (IIoT) or Industry 4.0. Industry 4.0

Advances in Computers, Volume 117
ISSN 0065-2458
https://doi.org/10.1016/bs.adcom.2019.10.010

dubbed I4.0 marks the fourth in the Industrial Revolution that focuses heavily on inter-connectivity, automation, autonomy, machine learning, and real-time data. By 2020, it is estimated that over 30 billion of the world's devices will be connected in some way—which is 20 billion more devices than today! The consistent capturing and transmitting of data among machines provide manufacturing companies with many growth opportunities. The IIoT is expected to transform how we live, work and play. The number one challenge faced by the Industrial IoT is security and privacy. If we cannot alleviate many of the security and privacy issues that impact the Industrial IoT, we will not be able to achieve its full potential. IoT and the trend toward greater connectivity means more data gathered from more places, in real time, to enable real-time decisions and increase revenue, productivity, and efficiency.

1. Introduction

The physical world is transformed into being digitized and make everything connected. An explosion of smart devices and technologies has allowed mankind to be in constant communication anywhere and any-time. The Internet of Things (IoT) is the network of physical objects—devices, instruments, vehicles, home appliances, buildings and other items embedded with electronics, circuits, software, sensors and network connectivity that enables these objects to collect and exchange data. The Internet of Things allows objects to be sensed and controlled remotely across existing network infrastructure, creating opportunities for more direct integration of the physical world into computer-based systems, and resulting in improved efficiency and accuracy (Fig. 1).

The efficiency through the use of hydropower, increased use of steam power and development of machine tools was achieved by the first industrial revolution; the second industrial revolution brought electricity and mass production of assembly lines; the automation using electronics and information technology was further accelerated by the third industrial revolution and recently the fourth revolution is emerging which is led by CPS technology to integrate the real world with the information era for future industrial advancement. Cyber physical systems (CPS) are an emerging discipline that involves engineered computing and communicating systems interfacing the physical world. Industrial Internet of Things is driving the fourth phase of the Industrial Revolution. It will accelerate the journey of Industries to a new level and transform economies, opening a new era of growth and competitiveness.

Expanding IoT capabilities and solutions enable new products and services that are transforming the way we work and play. Some of the emerging

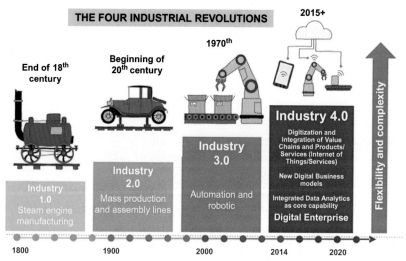

Fig. 1 The four industrial revolutions.

fields in the fourth industrial revolution include IoT, robotics, machine learning, artificial intelligence (AI), nanotechnology, quantum computing and biotechnology. In a smart and connected home, homeowners can have remote access to control appliances, security and surveillance systems, and much more—through their smartphones. In smart and connected cities across the world IIoTs are being deployed with the promise of saving energy, using limited resources more efficiently, and reducing carbon footprints to create a greener planet. In smart and connected automobiles, sensors are being implemented to increase driver and passenger safety and security; these regulations may be mandated in all new cars in the future. In almost every walk of life, the IoT seems to be changing our world. By 2020, it is estimated that over 30 billion of the world's devices will be connected in some way—which is 20 billion more devices than today!

The objective of this chapter is to provide readers with general concept of IIoT, the architecture and layers in IIoT, applications of IIoT with it and the challenges and opportunities provided by IoT.

2. Defining Industrial Internet of Things

The Industrial Internet of Things is connecting the physical world of sensors, devices and machines with the Internet and, by applying deep analytics through software, is turning massive data into powerful new insight and intelligence. The industrial Internet of Things (IIoT) refers to the extension and use of the Internet of Things (IoT) in industrial sectors and applications. With a strong focus on machine-to-machine (M2M) communication, big data, and machine learning, the IIoT enables industries and enterprises to have better efficiency and reliability in their operations.

IoT can be divided into three categories, based on usage and clients base:

Consumer IoT includes the connected devices such as smart cars, phones, watches, laptops, connected appliances, and entertainment systems.

Commercial IoT includes things like inventory controls, device trackers, and connected medical devices.

Industrial IoT covers such things as connected electric meters, waste water systems, flow gauges, pipeline monitors, manufacturing robots, and other types of connected industrial devices and systems (Fig. 2).

We are entering an era of profound transformation as the digital world is maximizing the efficiency of our most critical physical assets. We are experiencing incredible innovation around the internet as it accelerates the connection of objects not only with humans but with other objects. *IIoT applications* promise to bring immense value into our lives.

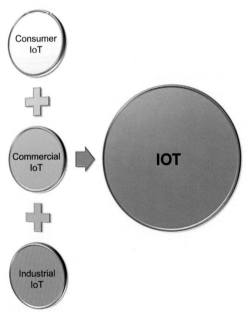

Fig. 2 Types of IoT.

2.1 History

Back in the day, coal miners would place a canary in a cage and carry the canary down into the mine shafts. The bird being more sensitive, would fall over, and would serve as an early indication that something toxic was in the air and it was not a safe place to mine, thus telling the miners to get out before they too fainted like the canary (Figs. 3 and 4).

Fig. 3 The canary in the coal mine. Image Courtesy of the Museum of Cannock Chase, Copyright Unknown.

Fig. 4 The canary in the IoT.

Today sensors mimic the canary in that they also "sensor" your surroundings and give off a warning that something is not quite right in your environment and that you need to take action. Fortunately for canary welfare, today's miners now use sophisticated infrared and catalytic "heat of combustion" sensors to detect poisonous and combustible gases. Hook these sensors up to the internet and now a safety officer in a remote location can monitor and even predict potential hazards before they have a chance to escalate.

The first telemetry system was rolled out in Chicago way back in 1912. It is said to have used telephone lines to monitor data from power plants. Telemetry expanded to weather monitoring in the 1930s, when a device known as a radiosonde became widely used to monitor weather conditions from balloons. In 1957 the Soviet Union launched Sputnik, and with it the Space Race. This has been the entry of aerospace telemetry that created the basis of our global satellite communications today.

Broad adoption of M2M technology began in the 1980S with wired connections to SCADA (supervisory control and data acquisition) on the factory floor and in home and business security systems. In the 1990s, M2M began moving toward wireless technologies. The concept of a network of smart devices was discussed as early as 1982, with a modified Coke machine at Carnegie Mellon University becoming the first internet-connected appliance, able to report its inventory and whether newly loaded drinks were cold.

2.2 Birth of IoT

The term "The Internet of Things" (IoT) was coined by Kevin Ashton in a presentation to Proctor & Gamble in 1999. He is a co-founder of MIT's Auto-ID Lab. He pioneered RFID (used in bar code detector) for the supply-chain management domain. In fact, already in 1926 we saw the first predictions of an Internet of Things. Back then, Nikola Tesla told Colliers Magazine the following in an interview: "When wireless is perfectly applied the whole earth will be converted into a huge brain, which in fact it is, all things being particles of a real and rhythmic whole.........and the instruments through which we shall be able to do this will be amazingly simple compared with our present telephone. A man will be able to carry one in his vest pocket." In the coming years we will see an explosion of devices connected to the Internet and together we are creating a smart planet.

Before the Internet was developed in 1969, Alan Turing already proposed the question whether machines can think, in his 1950 article

Computing Machinery and Intelligence. He stated that "...It can also be maintained that it is best to provide the machine with the best sense organs that money can buy, and then teach it to understand and speak English. This process could follow the normal teaching of a child." So, years before the first message was send across Internet, Alan Turing was already thinking about smart machines communicating with each other.

2.3 IoT timeline

To fully grasp the idea of IoT, we need to first figure out how it has been developing through time. We will skip all the "electricity invention" stage and build a more significant timeline. We will take a closer look at what IoT is and what it was to acquire a stronger understanding of its history.

In today's fast-paced world of technology, new software can reshape industries overnight. But while the Industrial Internet of Things (IIoT) is everywhere now—connecting millions of devices, machines, sensors, and systems throughout the world—it is anything but an overnight sensation. Over the past 50 years, technological milestones big and small—from large-scale system architecture breakthroughs to modest "Eureka" moments—have led to today's IIoT and are still informing predictions for tomorrow's industrial landscape.

While it's far from comprehensive, the timeline below should give you a general idea of where IoT has come from and where it's headed in the future. The road to the creation of the IIoT started in 1968, when engineer Dick Morley made one of the most important breakthroughs in manufacturing history. That year, Morley and a group of geek friends invented the programmable logic controller (PLC), which would eventually become irreplaceable in automating assembly lines and industrial robots in factories.

Today's infographic comes to us from Kepware, and it shows how these technological forces have emerged over time to make the IIoT possible (Fig. 5).

3. IIoT architectures

One of the most important challenges that companies face while opting for IIoT is the choice of architecture. Since the concept of IIoT itself is based on connecting devices in a network, the architecture plays a critical role. IIoT systems are often conceived as a layered modular architecture of digital technology.

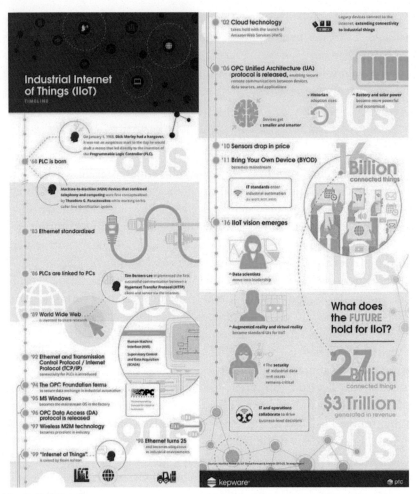

Fig. 5 IIoT timeline.

Different industries have different IIoT requirements. For example, the data output of an airplane taking off and landing is very different than that of a simple gasket manufacturing process. Since IIoT involves transmission of data, choosing the right IIoT architecture is therefore important. Successful digital transformation implementations are driven by a grounded business plan and an overall operational architecture. Designing Industrial IoT (IIoT) systems enforces new sets of architectural decisions on software/system architects. Let's look at the building blocks of IIoT:

Things. A "thing" is an object equipped with sensors that gather data which will be transferred over a network and actuators that allow things to act (for example, to switch on or off the light, to open or close a door, to increase or decrease engine rotation speed and more). This concept includes fridges, street lamps, buildings, vehicles, production machinery, rehabilitation equipment and everything else imaginable. Sensors are not in all cases physically attached to the things: sensors may need to monitor, for example, what happens in the closest environment to a thing

Gateways. Data goes from things to the cloud and vice versa through the gateways. A gateway provides connectivity between things and the cloud part of the IoT solution, enables data preprocessing and filtering before moving it to the cloud (to reduce the volume of data for detailed processing and storing) and transmits control commands going from the cloud to things. Things then execute commands using their actuators
Cloud gateway facilitates data compression and secure data transmission between field gateways and cloud IoT servers. It also ensures compatibility with various protocols and communicates with field gateways using different protocols depending on what protocol is supported by gateways. Streaming data processor ensures effective transition of input data to a data lake and control applications. No data can be occasionally lost or corrupted

Data Lake Storage

Data lake. A data lake is used for storing the data generated by connected devices in its natural format. Big data comes in "batches" or in "streams." When the data is needed for meaningful insights it's extracted from a data lake and loaded to a big data warehouse. *Big data warehouse* contains only cleaned, structured and matched data (compared to a data lake which contains all sorts of data generated by sensors). Also, data warehouse stores context information about things and sensors (for example, where sensors are installed) and the commands control applications send to things

Continued

—Cont'd

Data analytics. Data analysts can use data from the big data warehouse to find trends and gain actionable insights. When analyzed (and in many cases—visualized in schemes, diagrams, infographics) big data show, for example, the performance of devices, help identify inefficiencies and work out the ways to improve an IoT system (make it more reliable, more customer-oriented). Also, the correlations and patterns found manually can further contribute to creating algorithms for control applications *Machine learning, and the models ML generates.* With machine learning, there is an opportunity to create more precise and more efficient models for control applications. Models are regularly updated (for example, once in a week or once in a month) based on the historical data accumulated in a big data warehouse. When the applicability and efficiency of new models are tested and approved by data analysts, new models are used by control applications

In essence, IoT architecture is the system of numerous elements: sensors, protocols, actuators, cloud services, and layers. The IoT technology stack consists of three tiers: sensor devices, gateways, and the data center or cloud IoT platform (Fig. 6).

3.1 Edge tier

The *Edge tier* focuses on gathering information via variety of sensors. Because sensors are so tiny and inexpensive, they can be embedded in many different types of devices, including mobile computing devices, wearable technology, and autonomous machines and appliances. They capture information about the physical environment, such as humidity, light, pressure, vibration and chemistry. Standards-based wired and wireless networking protocols are used to transmit the data from the device to the gateway.

3.2 Platform tier

The *Platform Tier* also known as *Gateway Tier*, acts as an intermediary that facilitates communications, offloads processing functions and drives action. Because some sensors generate tens of thousands of data points per second, the gateway provides a place to preprocess the data locally before sending on to the data center/cloud tier. When data is aggregated at the gateway,

IIoT Network Reference Architecture

Three Functional Tiers

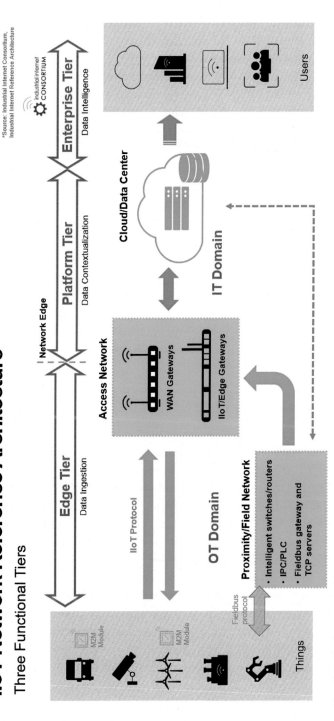

Fig. 6 IIoT reference architecture.

summarized and tactically analyzed, it can minimize the volume of unnecessary data forwarded on. Minimizing the amount of data can have a big impact on network transmission costs, especially over cellular networks. It also allows for critical business rules to be applied based on data coming in. The control tier is bidirectional. It can issue control information southbound, such as configuration changes. At the same time, it can respond to northbound device command-and-control requests, such as a security request for authentication.

3.3 Cloud tier

The *data center/cloud tier* performs large-scale data computation to produce insights that generate business value. It offers the back-end business analytics to execute complex event processing, such as analyzing the data to create and adapt business rules based on historical trends, and then disseminates the business rules downstream (southbound). It needs to scale both horizontally (to support an ever-growing number of connected devices) as well as vertically (to address a variety of different IoT solutions). Core functions of an IoT data center/cloud platform include connectivity and message routing, device management, data storage, event processing and analysis, and application integration and enablement.

4. Applications of IIoT

The IIoT will revolutionize manufacturing by enabling the acquisition and accessibility of far greater amounts of data, at far greater speeds, and far more efficiently than before. Several companies have started to implement the IIoT by leveraging intelligent, connected devices in their factories. The Industrial Internet of Things (IIoT) is driving strong demand for more data acquisition, communication, real-time analytics and data-driven decisions across a wide range of industrial verticals. The Industrial Internet of Things (IIoT) is no longer a futuristic idea for companies around the world. More and more, it's the way business is done.

Consumers are enthusiastic for smart homes, smart cars, smart gardens, smart kitchens and smart appliances, whether that intelligence is created through AI robotics or the Internet of Things. Enterprises are finding new business opportunities from the exabytes of edge device data, data that can be organized into simple and understandable ideas through AI.

In comparison to IoT, Internet of Everything (IoE) is the evolution of technology, business strategies, and human-to-technology engagement in ways that reinvent how organizations function. The IoE includes technology solutions that combine security, software-defined networking, unified communications, analytics, application-aware networking, database federation, and mobile experience. The Internet of Things is the key enabler of the IoE, as shown (Fig. 7).

Everyone agrees that the Internet of Things (IoT) has enormous economic potential. The vision: 50 billion connected devices delivering greater efficiency, productivity, safety, and comfort in every aspect of our daily life. From wearable devices and connected cars, to connected homes, smart cities, and industrial infrastructure, the power and potential of the IoT is generating excitement and spurring innovation in products, services, and business models.

Value creation in IoT comes from the ability to make real-time, data-driven decisions. This requires data capture, transmission, analysis, and storage. Data-capture applications are comprised of billions of connected intelligent and semi-intelligent "Things" that require M2M connectivity. Because these "Things" are deployed in various infrastructure locations like city parking lots, traffic lights, energy farms, etc., they must be remotely managed for code, data analytics, security, and business policies. Data transmission requires multiple levels of connectivity and relies on the ability to connect distributed IoT devices to a decision-management system. Data analysis requires more compute power and memory—specially to drive real-time decisions. Data storage needs are growing as well due to the growth of digital content and machine-generated content.

4.1 Smart manufacturing

A vision of tomorrow's manufacturing products finding their way independently through the production process. In intelligent factories machines and products communicate with each other cooperatively driving production raw materials and machines are interconnected within an Internet of Things. Factories with network machines and products are already in existence today in the future however these hitherto self-contained systems will be connected together in a comprehensive network all devices machines and materials will be duly equipped with sensors and communications technology and connect to each other these systems are known as cyber physical systems (CPS). The highlight is that they communicate with each other and control each other cooperatively. Industry 4.0 is based on the logic of cyber physical systems.

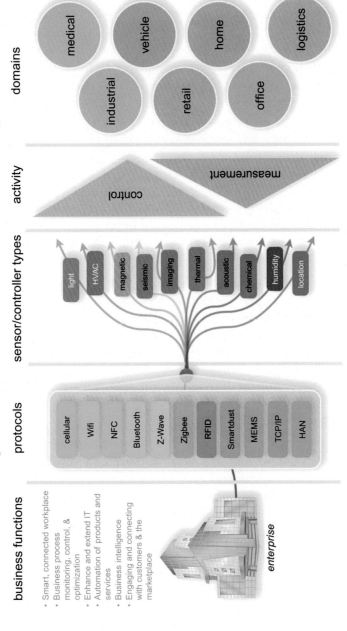

Fig. 7 Enterprise view of the Internet of Things.

Smart manufacturing allows factory managers to automatically collect and analyze data to make better-informed decisions and optimize production. Smart manufacturing is all about harnessing data; data will tell us "what to do" and "when to do it". It is a powerful disruptive force with the potential to restructure the current competitive landscape in manufacturing. Industrial manufacturing equipment is often subject to progressive wear. Monitoring and maintenance procedures are therefore required, however, they are difficult to implement autonomously and in situ. As they must monitor potentially inaccessible environments without interrupting near-continuous production, manufacturing processes are prime candidates for remote monitoring techniques. Sensor deployment in industrial operations is increasing rapidly due to significant reductions in sensor costs, advancements in sensing technology and the introduction of advanced analytical applications that can be used to extract and reveal insights in the data. This rise in the use of sensors has recently been labeled as the Industrial Internet of Things or (IIoT).

- IIoT incorporates machine learning and big data technology, harnessing the sensor data, machine-to-machine (M2M) communication and automation technologies that have existed in industrial settings for years.
- The driving philosophy behind the IIoT is that smart machines are better than humans at accurately, consistently capturing and communicating data.
- This data can enable companies to pick up on inefficiencies and problems sooner, saving time and money and supporting business intelligence efforts.
- In manufacturing specifically, IIoT holds great potential for quality control, sustainable and green practices, supply chain traceability and overall supply chain efficiency.

The data from sensors and machines is communicated to the Cloud by IoT connectivity solutions deployed in the factory. That data is analyzed and combined with contextual information and then shared with authorized stakeholders. IoT technology, leveraging both wired and wireless connectivity, enables this flow of data, providing the ability to remotely monitor and manage processes and change production plans quickly, in real time when needed. It greatly improves outcomes of manufacturing reducing waste, speeding production and improving yield and the quality of goods produced.

Smart manufacturing requires both horizontal and vertical data integration across the business. Vertical digitization may include manufacturing, procurement, supply chain, design product life cycle management logistics operations and quality all integrated for seamless flow of data. Horizontal digitalization may include data integration with suppliers, customers and key partners achieving integration requires upgrading replacing equipment networks and processes until you.

Harbor Research and Postscapes, describes the Internet of Things (IoT) as a digital nervous system for the world. Whole IoT implementation as a living, breathing entity that uses its senses to see, hear and feel the environment. If information derived from a finger says something is too hot, it needs that information to be delivered quickly to the brain, so the hand can be pulled back. Eyes and ears using cameras and microphones, along with sensory organs that can measure everything from temperature to pressure changes (Fig. 8).

Audio is the most apparent signal of mechanical failure. Most of the faults are signaled in this domain because of the movement of the components in the motor pumps creating friction. Microphones are listening for signs of wear—for variances to develop in the noises made by the machines—so that maintenance can be scheduled before anything breaks and causes downtime. Downtime, as you might imagine, is about the worst thing that can happen to a manufacturing facility. Rather than direct analysis of the time-series waveforms, audio signals were transformed into spectrograms and treated as pseudo-images. Analogous to pixels in images, the concept of an

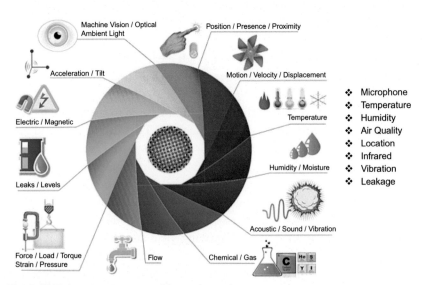

Fig. 8 Digital nervous system with super sensors.

"auxel" was used to represent the intensity value of a spectrogram at a specific time-frequency location. This image-like data format enabled representation of events with a diverse range of spatial features, such as edges; defined edge-like features in spectrograms such as high energy lines in the frequency domain are comparable to edges of an object in an image.

One of our key IIoT smart manufacturing in semiconductor is initiatives monitors wafer tool health by applying acoustic sensors to wafer processing machines to collect signals as wafers are polished. By comparing baseline sound fingerprints from a wafer's acoustic signals to sound fingerprints detected during polishing, we can recognize abnormal sounds and identify potentially problematic tool conditions (Fig. 9).

Sensors used in traditional manufacturing, like torque and pressure, are not sensitive to these defects. Acoustic monitoring system enables us to gather acoustic signals from the robotics in semiconductor tool and identify abnormal sound. Advanced signal processing techniques helps in extract features and apply machine learning techniques to differentiate the problematic polish sound fingerprint from the baseline sound fingerprint.

Another use case in sophisticated manufacturing is relative ease of adding online pump condition monitoring with today's wireless sensor technology, online monitoring can be installed quickly and inexpensively.

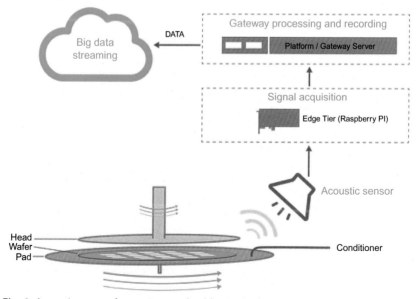

Fig. 9 Acoustic sensor for equipment health monitoring.

Cavitation monitoring is needed on high-head multistage pumps, as they cannot tolerate this condition, even for a brief time. Damage can occur before manual rounds discover the problem but can be detected sooner by continuously monitoring the pump discharge pressure for fluctuations with a wireless pressure transmitter.

Vibration monitoring detects many common causes of pump failure. Excessive motor and pump vibration can be caused by a failing concrete foundation or metal frame, shaft misalignment, impeller damage, pump or motor bearing wear, or coupling wear and cavitation. Increasing vibration commonly leads to seal failure and can result in expensive repairs, process upsets, reduced throughput, fines if hazardous material is leaked, and fire if the leaked material is flammable.

4.2 Smart agriculture

The global population is set to touch 9.6 billion by 2050. So, to feed this much population, the farming industry must embrace IoT. Against the challenges such as extreme weather conditions and rising climate change, and environmental impact resulting from intensive farming practices, the demand for more food has to be met. *Smart agriculture* based on IoT technologies will enable growers and farmers to reduce waste and enhance productivity ranging from the quantity of fertilizer utilized to the number of journeys the farm vehicles have made. Technologies and IoT have the potential to transform agriculture in many aspects. Namely, there are five ways IoT can improve agriculture:

- Data, tons of data, collected by smart agriculture sensors, e.g., weather conditions, soil quality, crop's growth progress or cattle's health. This data can be used to track the state of your business in general as well as staff performance, equipment efficiency, etc.
- Better control over the internal processes and, as a result, lower production risks. The ability to foresee the output of your production allows you to plan for better product distribution. If you know exactly how much crops you are going to harvest, you can make sure your product won't lie around unsold.
- Cost management and waste reduction thanks to the increased control over the production. Being able to see any anomalies in the crop growth or livestock health, you will be able to mitigate the risks of losing your yield.
- Increased business efficiency through process automation. By using smart devices, you can automate multiple processes across your production cycle, e.g., irrigation, fertilizing, or pest control.

- Enhanced product quality and volumes. Achieve better control over the production process and maintain higher standards of crop quality and growth capacity through automation.

4.3 Smart cities

Smart city deployment is multidimensional. In any major urban center, there are several use cases, and they can vary depending on the size of the city and local jurisdictional control. The rise of the smart home is one way that the IoT is changing things for people around the world—homes filled with devices that can communicate with one another, with people living in the home, and even with outside third parties (think a refrigerator automatically contacting a grocery store when key items like milk or eggs run low to order more).

There are quite a few ways in which this sort of technology is being utilized today around the globe, including the following:

- Sensors built into bridges to sense things like degradation and the effect of seismic forces at work.
- Sensors built into roadways to sense things like subsidence and wear and tear, as well as traffic flow.
- Sensors built into buildings to sense things like the force of wind, foundation subsidence, seismic activity, and more.
- Sensors within the interior of buildings to sense the presence of people within rooms, and thereby control the use of lighting, heating and air, and other systems to limit energy expenditure when it is unnecessary.
- Sensors at entryways to provide facial recognition for better security within apartment buildings, commercial buildings, government offices, and more.

4.4 Smart home

Smart home technology, also often referred to as home automation or domotics (from the Latin "domus" meaning home), provides homeowners security, comfort, convenience and energy efficiency by allowing them to control smart devices, often by a smart home app on their smartphone or other networked device. A part of the Internet of Things (IoT), smart home systems and devices often operate together, sharing consumer usage data among themselves and automating actions based on the homeowners' preferences.

Smart TVs connect to the internet to access content through applications, such as on-demand video and music. Some smart TVs also include voice or gesture recognition. In addition to being able to be controlled remotely and customized, smart lighting systems, such as Hue from

Philips Lighting Holding B.V., can detect when occupants are in the room and adjust lighting as needed. Smart lightbulbs can also regulate themselves based on daylight availability.

Smart thermostats, such as Nest from Nest Labs Inc., come with integrated Wi-Fi, allowing users to schedule, monitor and remotely control home temperatures. These devices also learn homeowners' behaviors and automatically modify settings to provide residents with maximum comfort and efficiency. Smart thermostats can also report energy use and remind users to change filters, among other things.

Using smart locks and garage-door openers, users can grant or deny access to visitors. Smart locks can also detect when residents are near and unlock the doors for them. With smart security cameras, residents can monitor their homes when they are away or on vacation. Smart motion sensors are also able to identify the difference between residents, visitors, pets and burglars, and can notify authorities if suspicious behavior is detected.

Pet care can be automated with connected feeders. Houseplants and lawns can be watered by way of connected timers. Kitchen appliances of all sorts are available, including smart coffee makers that can brew you a fresh cup as soon as your alarm goes off; smart refrigerators that keep track of expiration dates, make shopping lists or even create recipes based on ingredients currently on hand; slower cookers and toasters; and, in the laundry room, washing machines and dryers.

Household system monitors may, for example, sense an electric surge and turn off appliances or sense water failures or freezing pipes and turn off the water so there isn't a flood in your basement.

4.5 Smart healthcare

AI in healthcare has been focused on reducing costs (always a requirement) and improving patient outcomes with the things that AI does best: finding patterns for improved diagnoses and identifying new treatments. AI technology is currently focused on applying deep learning algorithms to detect and highlight abnormalities in medical imaging. AI is also assisting patients with chatbots to schedule appointments, make billing more efficient, or just providing basic medical information.

Precision medicine is about diagnosing illness through machines using AI trends and defining the treatment plan; understanding the associated genomics and predicting the probability of success of the treatment; and ultimately zooming in on the right treatment and monitoring it. Maximum accuracy requires algorithms that are trained separately for every condition and disease, which is costly and labor intensive, so much of this is in the development process now and won't be commercialized for at least a few years. The FDA and other countries' regulatory oversight must also approve any systems and doctors must integrate it into their practices.

Per Forbes in 2018: "The accuracy gap between the human and digital eye is expected to widen further, and soon." The healthcare field for AI is already helping to advance diagnostics in radiology, pathology, dermatology and ophthalmology (by analyzing veins in the eye). Medical professionals are able to more efficiently address lung cancer, melanoma, cardiovascular disease and the effects of diabetes on vision.

Precision health is about monitoring your health every day, every second, outside of treatment centers and hospitals. These applications, once created, approved and launched, will be creating massive amounts of data to store and analyze.

4.6 Smart transportation

Our lives run on transportation. It gets us to work in the mornings, delivers fresh food to our grocery stores, and encourages us to travel to far corners of our wide world. IoT (Internet of Things) is a disruptive digital innovation with the capacity to enhance so many aspects in our everyday lives, including transportation.

Ford Motor Company didn't invent the automobile, but with the introduction of the Model T in 1908, they began a revolution that changed the world. For all their importance, cars really haven't changed that much over the last century in terms of purpose or functionality. Combined with the advancements triggered by the Internet of Things (IoT), the automotive industry is now on the cusp of a second revolution that could be even bigger than Ford's Model T—a technology revolution.

In the automotive market, futuristic technologies are coming to life more quickly than anyone imagined. Automobiles are poised to become an integral part of the IoT. IoT technologies like cameras that gather and process data to enable the driver and vehicle to respond more quickly and efficiently have already been integrated into vehicles and are resulting in a safer driving experience. As Advanced Driver Assistance Systems (ADAS) are implemented in many vehicles, the rapid learning of these systems is forming the foundation for fully autonomous vehicles.

While fully autonomous vehicles are not yet common on our roads, there are already five systems within the vehicle that will help connect it to the IoT and lay the groundwork for full autonomy: ADAS, infotainment, instrument cluster, powertrain and communications/telematics. These cars will have multiple cameras, sensors, radars, LiDAR on them that will be creating data through computer vision to facilitate safe navigation. All this data must be processed fast, millions of decisions made within a split second inside the car, as well as connecting with the cloud to realize the full potential of autonomous vehicles of the future.

Autonomous vehicles are going to drive network growth to levels that were unimaginable in the past. During their developer conference in August of 2016, Intel projected that by 2020 every autonomous car would create and send 4000GB of data daily to networked cloud systems that manage communication among IoT devices. This is five times the data created by the average individual today. Connected airplanes and factories (to name just a few) will create even greater data requirements.

We've already started to see the incorporation of IoT in transportation. They can be as small scale as a vehicle-to-person communication (Your Lyft is two minutes away!) and as large scale as monitoring shipping logistics of a global company. Let's discuss IoT applications in transportation that are revolutionizing the industry.

5. Securing Internet of Things

By now most people have heard about the onslaught of Internet-connected devices that is expected to become part of our lives over the next several years. This connectivity will bring unprecedented productivity, new business models, and efficiencies never thought possible. It will also offer tantalizing targets for a new breed of criminals and other bad actors looking to break into virtual back doors that never existed before the castle's moat became so shallow. With this foresight, it's easy to see why tomorrow's interconnected devices must be designed from the ground up with security in mind.

Unsecured IoT devices are real concern given that more than 25% of cyberattacks will involve IoT devices. Putting policies in place will protect people, while enabling this transition. Gartner estimates that 26 billion units would be added to the internet by 2020 and this is exclusive of the PCs, laptops and smartphones. It is pretty critical to secure all these systems that are well on the way of becoming essential parts of our lives (Fig. 10).

IoT devices are meant to be tiny, so objective specific. They do what they are intended to do–collect data, rely it or access connectivity. This means that additional hardware for security at that size are yet nascent. Like a domino effect, one single vulnerability found in a single IoT device in the network could lead to massive security nightmare to the entire network of such connected devices.

We will see an increase in this and the advent of contextual data sharing, and autonomous machine actions based on that information, the IoT is the

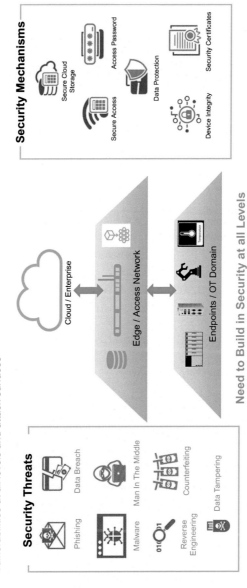

Fig. 10 Security threats in the IIoT network.

allocation of a virtual presence to a physical object, as it develops, these virtual presences will begin to interact and exchange contextual information, (and) the devices will make decisions based on this contextual device. This will lead to very physical threats, around national infrastructure, possessions (for example, cars and homes), environment, power, water and food supply, etc.

The IIoT spans multiple industries including power and energy, oil and gas, manufacturing, chemical plants, healthcare, and aviation. The widespread global deployment of IIoT infrastructure has provided an attractive platform for attackers to digitally infiltrated the devices. Internet of Things cannot simply be left to become the "Internet of Threats."

The threats under IoT are broadly classified under these three areas—safety, security, and privacy of data.

5.1 Safety

IIoT brings the ever-increasing digital risks associated with cybersecurity into physical spaces, creating a vast array of new vulnerabilities including threats to public safety, physical harm and catastrophic systemic attacks on commonly shared public infrastructure. When an oil pipeline was hacked in Turkey causing an explosion and 30,000 barrels of spilled oil, the cyber attackers negated the existing safety system to shut down alarms, cut off communications and super-pressurize crude oil in the line. A regional water supplier experienced a cyber-security breach that not only compromised customer data, but caused unexplained valve and duct movements, including manipulation of PLCs that managed water treatment and public safety. There's a possibility of connected equipment malfunctioning and causing harm to people nearby.

Additionally, if hackers gain access to connected machinery and make it behave in ways that are undetectable for a while, they could cause the equipment to make defective equipment. That possibility also compromises safety. In industrial production, your safety and security programs are inextricably linked. Many of our customers are tapping IIoT technology to remotely access production machinery, allow wireless access to pumping stations, or connect plant-floor equipment to the IT infrastructure. This is the future. This is how they can realize improved asset utilization, faster time to market, and lower total cost of ownership. However, greater connectivity can increase security risks that will impact safety. This is where better enterprise risk management is important.

Users' increased demands for consistent, uninterrupted operation of these legacy systems. This is especially true for infrastructures, such as power grids, pipelines, or water systems, where downtime can be catastrophic.

Cyber threat landscape

IoT verticals

Health care | Life sciences
- Patient care
- Remote diagnostics
- Bio wearables
- Food sensors
- Equipment monitoring
- Elderly monitoring

Smart home | Consumer
- Wearables
- Smart thermostat
- Pet feeding
- Smoke alarm
- Refrigerator
- Washer
- Home security

Infrastructure | Cities
- HVAC
- Smart city
- Smart building
- Waste management
- Temperature control
- Electric vehicle

Transport | Urban mobility
- Traffic routing
- Telematics
- Smart parking
- Public transport

Industrial systems | Sensors
- Speed
- Temperature
- Flow
- Motor
- Valve
- Pressure
- Light
- Position
- Fan
- Heat/cool
- Robot arm

Ecosystem

Carrier/Network access · Wi-Fi · Software (Platforms/databases, Embedded OS, Middleware, Protocols) · RFID · Connected things · Consumer IP/Wi-Fi · Enterprise IP/Wi-Fi · Enterprise services · Enterprise IP networks · Home automation wireless protocols · Cloud services · Public IP networks · Orchestration services · Product developers · System integrators · Support teams · Analytics & instrumentation services · Hardware (Microprocessors, Sensors, Mobile chips, Actuators, Integrated kits) · Hardware manufacturers · NFC · Bluetooth

The cyber risk landscape is inexhaustibly complex and ever changing. This figure provides a broad framework for identifying and managing a much wider range of risks arising from IoT implementations.

Source: Deloitte & Touche LLP

Cyber security drivers

Security controls
- Data protection
- End-point protection
- Trust and safety
- Physical security
- Security monitoring and analytics
- Security resiliency
- Regulatory compliance
- Identity and access management

Threats & vulnerabilities
- Counterfeiting
- Denial of service
- Eavesdropping
- Buffer overflow
- Malicious modification
- Password-based attacks
- Man-in-the-middle attacks
- Phishing

The most important requirement for collections of devices is that they guarantee physical safety and personal security. There are significant benefits in having all vehicles and related infrastructure connected to the IoT; however, there is also significant risk. How can we mitigate the risks of future cyber-attacks as vehicles become more connected? Hacking a car and stealing private information seems trivial when compared to the threat of having someone take control of the car from the outside and putting lives in danger.

The next technology step that will provide even more safety on the road is the creation of an integrated system network to connect vehicle systems together via the IoT to communicate and interact in real time with their surroundings and ensure maximum safety for all involved.

5.2 Security

Securing IoT devices is essential, not only to maintain data integrity, but to also protect against attacks that can impact the reliability of devices. As devices can send large amounts of sensitive data through the Internet and end users are empowered to directly control a device, the security of "things" must permeate every layer of the solution.

One of the core problems with the increasing number of IoT devices is the increased complexity that is required to operate them securely. This increased complexity creates security challenges far beyond the difficult challenges' individuals face just securing a single device. We highlight some of the negative trends that smart devices and collections of devices cause, and we argue that issues related to security, physical safety, privacy, and usability are tightly interconnected and solutions that address all four simultaneously are needed. Tight safety and security standards for individual devices based on existing technology are needed. Likewise, research that determines the best way for individuals to confidently manage collections of devices must guide the future deployments of such systems.

The industry needs is the equivalent of a flu shot to protect its embedded devices. Each of the future nodes within the connected castle needs to be immunized against attacks before they can happen. This can only happen with more robust security capabilities within the memory device itself. Doing so will not only protect the computing device on the network, but at all points prior to the network, from its inception through the supply chain.

Better memory security will in no way obviate the need for the sound practices of code signing and measurement. After all, good security is always based on layers of security. The ultimate goal, however, should be that by protecting the memory along with the system logic, the computing device as a whole will become much more resilient to attack.

Instead, the future lies in a mix of cloud and fog computing, where some computing happens at the edge on devices in the field and some occurs in the cloud. Initial data will be processed in the field, anomalies and change data will be sent to the cloud for analysis, and on-site systems will provide continuous insight on system health to highlight potential areas for improvement.

While threats will always exist with the IoT as they do with other technology endeavors, it is possible to bolster the security of IoT environments using security tools such as data encryption, strong user authentication, resilient coding and standardized and tested APIs that react in a predictable manner.

Security needs to be built in as the foundation of IoT systems, with rigorous validity checks, authentication, data verification, and all the data needs to be encrypted. At the application level, software development organizations need to be better at writing code that is stable, resilient and trustworthy, with better code development standards, training, threat analysis and testing. As systems interact with each other, it's essential to have an agreed interoperability standard, which safe and valid. Without a solid bottom-top structure we will create more threats with every device added to the IoT. What we need is a secure and safe IoT with privacy protected, tough trade off but not impossible.

5.3 Privacy

Privacy is challenging to understand and guarantee in a world where more and more smart devices collect data, share it, and monetize it. The fundamentally different *privacy* vulnerabilities may well be an even bigger threat to the success of the IoT. The data collected by IoT devices are increasingly sounding the alarm that IoT users should be paying attention to what happens to the data collected by IoT devices. These devices are placed on the Internet for convenience, creating a significant public nuisance at the minimum, in some cases real infringement on privacy, and in the extreme are used to create broad Internet attacks.

Concerning sensitive personal data like location and movements. This situation raises concerns around maintaining privacy of the personal information. Many embedded connected devices are being designed today with the equivalent of the PC's signed BIOS and integrity measurement. How many of these devices will be able to be remotely and centrally verified that they have not been compromised before they come alive on the Internet? And when one considers the actual brains and nervous system of each connected device, the code which it runs, how exactly is this code protected?

One of the key tenants set forth by the Trusted Computing Group and most others in the security industry is the concept of "trusted" or "measured" boot. The idea being that a computing device at its core is really the software in which it is executing so this code must be verified to be good. At the earliest phase of the device's boot, its code will measure itself and then measure other code modules before transferring control to them. If each subsequent measurement is deemed correct and therefore trusted, then the device progresses through its boot process in a trusted manner and eventually becomes a trusted connected device.

So, to answer a question previously set forth, measurement of code before it is executed is the industry's answer to code protection. The problem with this answer is that it's very much like getting sick and confirming the origin of your ailment with a throat swab or blood test. Although it's nice to know that you are in fact sick and medicine exists to improve your condition, you can't help but think that you would rather have prevented the whole ordeal in the first place.

We have outlined some of the implications of these changes through a discussion of use-case scenarios and the dimensions of safety, security, and privacy. We believe that changes are happening with such speed and the level of risk and uncertainty is sufficiently high that investment in research that helps mitigate potential problems should be prioritized. The potential benefit to human lives, our national interests, and the economy is sufficient to warrant substantial research investments in making the technology as beneficial as possible.

Regulatory standards for data markets are missing especially for data brokers; they are companies that sell data collected from various sources. Even though data appear to be the currency of the IoT, there is a lack of transparency about; who gets access to data and how those data are used to develop products or services and sold to advertisers and third parties. There is a need for clear guidelines on the retention, use, and security of the data including metadata (the data that describe other data).

6. Challenges and opportunities

The introduction of the Internet of Things (IoT) into all kinds of businesses and industries presents a magnitude of benefits and potential challenges. Some of the most pressing challenges and concerns emerge from the adoption of IIoT. The challenges are:

Security vulnerabilities (privacy, sabotage, denial of service): Regular hacking of high-profile targets keeps this danger constantly in the back of our minds. Obviously, the consequences of sabotage and denial of service could be far more serious than a compromise of privacy. Changing the mix ratio of disinfectants at a water treatment plant or stopping the cooling system at a nuclear power plant could potentially place a whole city in immediate danger.

Connectivity: Connecting so many devices will be one of the biggest challenges of the future of IoT, and it will defy the very structure of current communication models and the underlying technologies [1]. At present we rely on the centralized, server/client paradigm to authenticate, authorize and connect different nodes in a network.

Regulatory and legal issues: This applies mainly to medical devices, banking, insurance, infrastructure equipment, manufacturing equipment, and in particular, pharmaceutical and food related equipment. Today, this mean complying with laws such as CFR 21 part 11, HIPAA, Directive 95/46/EC and GAMP 5, etc. This adds to the time and cost needed to bring these products onto the market.

Determinism of the network: This is important for almost all areas where IoT can be used, such as in control applications, security, manufacturing, transport, general infrastructure, and medical devices. The use of the cloud currently imposes a delay of about 200 milliseconds or more. This is fine for most applications, but not for security or other applications that require a rapid, almost immediate, response. A trigger from a security monitoring system received 5 s later could be too late.

Lack of a common architecture and standardization: Continuous fragmentation in the implementation of IoT will decrease the value and increase the cost to the end users. Currently, aside of the products mentioned above, there are also Google's Brillo and Weave, AllJoyn, Higgns, to name but a few. Most of these products target very specific sectors. Some the causes of this fragmentation are security and privacy fears (privacy through obfuscation and the fear of "not invented here"), jostling for market dominance, trying to avoid issues with competitors' intellectual property, and the current lack of clear leadership in this area.

Scalability: This is currently not much of an issue, but it is bound to become an issue mainly in relations to generic consumer cloud as the number of devices in operation rises. This will increase the data bandwidth needed and the time needed for verifying transactions.

Limitations of the available sensors: Fundamental sensor types, such as temperature, light, motion, sound, color, radar, laser scanner, echography and X-ray, are already quite performant. Furthermore, recent advances in microelectronics, coupled with advances in solid state sensors, will make the bare sensors less of an issue in the future. The challenge will be in making them more discriminating in crowded, noisy and more complex environments. The application of algorithms that are similar to fuzzy logic promises to make this less of an issue in the future.

Dense and durable off-grid power sources: While Ethernet, WIFI, 3G and Bluetooth have been able to solve most connectivity issues by accommodating the various devices' form factors, the limitations of battery life still remain. Most smartphones still need to be charged every day, and most sensors still need regular battery changes or connection to the grid. It would make a difference if power could be broadcasted wirelessly to such devices from a distance, or if power sources that can last for at least a year can be integrated into the sensors.

IIoT is the foundation of a digital transformation that will present new opportunities for manufacturers and redefine future business models. Some of the opportunities are:

Machine-as-a-service: Machine–as–a–service (MaaS) can be compared to energy performance contracting. Implementing technologies that allow a machine builder to understand the exact performance of the machines they build (including how much those machines can manufacture) is a revolutionary new opportunity to monetize industrial machines.

Predictive maintenance: The benefits from such IoT predictive maintenance transformation are significant—equipment downtime is reduced, and on–site service calls can be both scheduled more efficiently and made more productive.

Asset performance management: Through either remote or on premises monitoring, sensor data supplementing existing operational sensor-based data ecosystems can enhance real-time situational awareness, close information gaps and operationalize assets that were previously unmonitored or monitored manually. IIoT broadens the scope of asset data sources to heighten texture and resolution of existing information. Industries can use information to streamline maintenance costs, improve process efficiency and increase asset availability.

Improved planning and productivity: IIoT enables industries to monitor the physical state or location of people as well as mobile or geographically dispersed assets. When combined with process or asset data, industries can operationalize data sources that surround core industrial machinery to enhance safety, field force and operational efficiencies.

Data driven communication: IIoT sensors and devices produce data and information relevant to broader ecosystems. Sharing data across these ecosystems will transform communication between customers, businesses and across larger ecosystems. Secure data exchange across these broader data environments will create new business models and associated revenue streams through opportunities such as expanded service offerings, establishing data-driven collaboration and partnerships.

7. Future of IIoT

The future of IIoT has the potential to be limitless. The rise of industrial IoT will soon bring the factory of the future to reality. Advances to the industrial internet will be accelerated through increased network agility, integrated artificial intelligence (AI) and the capacity to deploy, automate, orchestrate and secure diverse use cases at hyperscale. An exciting wave of future IoT applications will emerge, brought to life through intuitive human to machine interactivity. Human 4.0 will allow humans to interact in real time over great distances—both with each other and with machines—and have similar sensory experiences to those that they experience locally. This will enable new opportunities within remote learning, surgery and repair. Immersive mixed reality applications have the potential to become the next platform after mobile—realized through 3D audio and haptic sensations and becoming our main interface to the real world. Bringing future IoT to life will require close synergy between the IoT- and network platforms (Fig. 11).

The Ignition IIoT solution greatly improves connectivity, efficiency, scalability, time savings, and cost savings for industrial organizations. It can unite the people and systems on the plant floor with those at the enterprise level. It can also allow enterprises to get the most value from their

Conversational Systems

Super Computing

Blockchain (Distributed Ledgers)

Big Data/Advanced Analytics

Virtual Reality and Augmented Reality

Digital Health

Intelligent Things

Adaptive Security Architecture

Artificial Intelligence and Advanced machine Learning

Digital Platforms

Fig. 11 Future of IIoT.

system without being constrained by technological and economic limitations. For these reasons and more, Ignition offers the ideal platform for bringing the power of the IIoT into your enterprise.

To gain the upper-edge on these looming forces, executing on an IIoT strategy is becoming invaluable in industrial enterprises. These adopters will benefit from extensive IIoT solutions quickly solving their business needs of today and capable of scaling to meet future requirements of tomorrow. The competencies of IIoT providers will become more apparent as these projects grow and these shifting requirements need solutions from providers with the right mix of technology, domain knowledge, and a partnership ecosystem to fill the gaps. Successful IIoT deployments in manufacturing and operations functions will continue to span across organizational hierarchies enabling synchronized operations with IIoT touchpoints ranging from the CXO to front-line worker.

Myriad global forces are causing organizations to become more "digital" at a rapid pace to avoid disruption, become more efficient, and capitalize on new opportunities. This digital transformation means different things to different industries. In manufacturing it comes back to operational efficiencies and becoming more flexible and agile.

The Industrial Internet of Things will drastically change the future, not just for industrial systems, but also. From wrist watches to smart appliances, we can already see this trend come to fruition. For the many people involved. If we can achieve the full potential of the Industrial IoT vision, many people will have an opportunity to better their careers and standards of living because the full potential of this vision will lead to countless value creation opportunities. This always happens when new "revolutions" get set into motion.

The full potential of the Industrial IoT will lead to smart power grids, smart healthcare, smart logistics, smart diagnostics, and numerous other "smart" paradigms. Material handling, manufacturing, product distribution and supply chain management will all be automated to a degree in the years to come. In the future, experts suggest that industrial IoT will enhance production levels even further and become the driving force behind various types of innovation (including the utilization of innovative fuels). It will enable manufacturing ecosystems driven by smart systems that have autonomic self-* properties such as self-configuration, self-monitoring, and self-healing. Think of a "Terminator" style technology. Instead, this is technology that will allow us to achieve unprecedented levels of operational efficiencies and accelerated growth in productivity.

While an array of adoption challenges will still have to be overcome, predictive analysis suggests that the world will have 50 billion connected devices by 2020. It would be a pity for such a massive network to remain unutilized in attempts to enhance industrial processes. Remember that industrial IoT is not about smart product development. Rather, it will help for a higher level of efficiency and predictive rather than reactionary interventions—a main problem industries across the world are struggling with today.

8. Conclusion

The IIoT is expected to transform how we live, work and play. The Industrial Internet of Things can have a bright and shiny future. From factory automation and automotive connectivity to wearable body sensors and home appliances, the IIoT is set to touch every facet of our lives. We will "author" our life with networks around us that constantly change and evolve based on our surroundings and inputs from other systems. Autonomous cars with IIoT makes our lives safer with cars that sense each other to avoid accidents. It will make our lives greener with lighting systems that adjust based on the amount of daylight from windows. It will make our lives healthier with wearables that can detect heart attacks and strokes before they happen.

The future of IoT is virtually unlimited due to advances in technology and consumers' desire to integrate devices such as smart phones with household machines. Wi-Fi has made it possible to connect people and machines on land, in the air and at sea. With so much data traveling from device to device, security in technology will be required to grow just as fast as connectivity in order to keep up with demands. Governments will undoubtable face tough decisions as to how far the private the sector is allowed to go in terms of robotics and information sharing.

It is critical that both technology companies and governments keep in ethics in mind as we approach the fourth Industrial Revolution. The number one challenge faced by the Industrial IoT is security and privacy. If we cannot alleviate many of the security and privacy issues that impact the Industrial IoT, we will not be able to achieve its full potential. IoT and the trend toward greater connectivity means more data gathered from more places, in real time, to enable real-time decisions and increase revenue, productivity, and efficiency. The possibilities are exciting, productivity will increase, and amazing things will come by connecting the world. There is a long road ahead to the IoT of 2020. But one thing is for sure, it is going to be amazing.

References

[1] Dano M: *Sierra: 2G M2M Sales for Long-Term Apps in the U.S. are Over*, 2013, FierceWireless.

Further reading

[2] ABIresearch: *More Than 30 Billion Devices Will Wirelessly Connect to the Internet of Everything in 2020*, 2013, ABIresearch. https://www.cisco.com/c/dam/en/us/products/collateral/se/internet-of-things/at-a-glance-c45-731471.pdf.

[3] Infonetics Research: *Mobile M2M Modules, Biannual Worldwide and Regional Market Size and Forecasts*, first ed., 2013, Infonetics Research. https://www.gartner.com/imagesrv/books/iot/iotEbook_digital.pdf.

[4] G. Laput, Y. Zhang C. Harrison: Synthetic sensors: towards general-purpose sensing, *CHI '17- Proceedings of the 2017 CHI Conference on Human Factors in Computing Systems*.

[5] Jones E: *Unlock the Processing Power of Wireless Modules*, 2013, Sierra Wireless.

[6] https://www.trendmicro.com/vinfo/us/security/definition/industrial-internet-of-things-iiot.

[7] https://arstechnica.com/information-technology/2019/06/manufacturing-memory-means-scribing-silicon-in-a-sea-of-sensors/.

[8] Schneider S: The industrial internet of things (IIoT). In Geng H, editor: *Internet of Things and Data Analytics Handbook*, Hoboken, NJ, USA, 2017, John Wiley & Sons, Inc. https://doi.org/10.1002/9781119173601.ch3.

[9] Research B: *M2 M Sector Map, 2014*. Available: http://www.beechamresearch.com/download.aspx?id=18.

[10] Jeschke S, Brecher C, Meisen T, Özdemir D, Eschert T: Industrial internet of things and cyber manufacturing systems. In *Industrial Internet of Things*, Cham, 2017, Springer, pp 3–19. https://doi.org/10.1007/978-3-319-42559-7_1. 2018.

[11] CyPhERS: Characteristics, Capabilities, Potential Applications of Cyber-physical Systems: A Preliminary Analysis, 2013. http://www.cyphers.eu/sites/default/files/D2.1.pdf. (Accessed 18 March 2018).

[12] Beecham Research: *M2 M Sector Map, 2014*. Available, http://www.beechamresearch.com/download.aspx?id=18.

[13] Department for Business, Energy & Industrial Strategy: *Made Smarter Review, 2017*, p 51. Fig. 16. Available, https://www.gov.uk/government/publications/made-smarter-review.

[14] Porter ME, Heppelmann JE: How smart, connected products are transforming competition: spotlight on managing the internet of things, *Harv Bus Rev* 92(11):64–88, 2014.

[15] Ågerfalk P, Levina N, Kien SS, editors: *Proceedings of the International Conference on Information Systems – Digital Innovation at the Crossroads, ICIS*; 2016.

[16] Püschel L, Roeglinger M, Schlott H: What's in a smart thing? Development of a multi-layer taxonomy. In Ågerfalk P, Levina N, Kien SS, editors: *Proceedings of the International Conference on Information Systems–Digital Innovation at the Crossroads, ICIS 2016, vol. 2016, Association for Information Systems, Dublin, Ireland, 2016 December*, 2016, pp 11–14.

[17] OWASP: IoT Framework Assessment, In *Internet of Things (IoT) Project*, 2018. Available. https://www.owasp.org/index.php/IoT_Framework_Assessment.

[18] Bradicich T: *The Intelligent Edge: What It Is, What It's Not, and Why It's Useful*, 2017, Hewlett Packard Enterprise Available, https://www.hpe.com/us/en/insights/articles/the-intelligent-edge-what-it-is-what-its-not-and-why-its-useful-1704.html.

[19] F. Bonomi, R. Milito, J. Zhu S. Addepalli: *Fog computing and its role in the internet of things*, https://conferences.sigcomm.org/sigcomm/2012/paper/mcc/p13.pdf.

[20] Bonomi F, Milito R, Zhu J, Addepalli S: Fog computing and its role in the internet of things. In *Proc SIGCOMM*, 2012.

[21] Lee EA: Cyber physical systems: design challenges. In *2008 11th IEEE International Symposium on Object and Component-Oriented Real-Time Distributed Computing (ISORC)*, 2008, pp 363–369.

[22] Breivold HP: *A Survey and Analysis of Reference Architectures for the Internet-of Things*, 2017, ICSEA, p 143.

[23] A. Al-Fuqaha, S. Sorour, M. Guizani and M. Mohammadi, *A Survey on Deep Learning for IoT Big Data and Streaming Analytics*, https://arxiv.org/pdf/1712.04301.pdf.

About the author

Dr. Sathyan Munirathinam is a Data Scientist at a largest Semiconductor Manufacturing Company. Working on Machine Learning / Deep Learning / IoT delivering innovative and cutting-edge AI and Machine Learning to the enterprise and developing tomorrow's advances in AI. Sathyan Munirathinam has a PhD in Machine Learning, with over 20 years in the Business Intelligence. He has authored many papers and involved in numerous international artificial intelligence and data mining research activities and conferences. His research interests include artificial intelligence, equipment health monitoring, IoT and big data analysis, statistical machine learning and data mining, ubiquitous computing, and human computer interaction.He holds a Master of Science from Illinois Institute of Technology, Chicago. He is a Certified Business Intelligence Professional.

CHAPTER SEVEN

The growing role of integrated and insightful big and real-time data analytics platforms

Indrakumari Ranganathan[a], Poongodi Thangamuthu[a], Suresh Palanimuthu[b], Balamurugan Balusamy[a]
[a]School of Computing Science and Engineering, Galgotias University, Greater Noida, India
[b]School of Mechanical Engineering, Galgotias University, Greater Noida, India

Contents

Abstract

Digitization era is altering several industries which include the way in which the data is analyzed and it is inferred that about 2.7 Zettabytes of data exist in the digital world today. By 2020 the data generated per second for every human being will approximate amount

165

to 1.7 megabytes and the volume of data would double every 2 years thus reach the 40 ZB point by 2020. Interactive Data Corporation (IDC) estimated that by the end of year 2020, the e-commerce transactions B2B and B2C will hit 450 billion per day on the internet.

The advent of Big and real time Data has triggered disruptive changes in many fields and the exploding volume of different sources of data like heterogeneous data, data integration, spatio-temporal correlation of data, batch analytics and real-time analytics, data sharing, semantic interoperability requires the development of a scalable platform that can fuse multiple data layers to handles the data intelligently.

In Big Data approaches, the challenge is not anymore to collect the data, but to draw valuable conclusions by properly analyzing them. The growth in Unstructured Data generated by business is irrefutable and they are under more pressure to preserve it for longer periods of time. To be clear, exploiting the collected data has been always considered by practitioners and researchers, but the huge velocity, heterogeneity and enormity of massive stream of real-time data shove the limits of the current storage, management and processing capabilities.

Admittedly, the traditional method of Extract, Transform and Load (ETL) are challenged and cannot be applied on the emerging opportunistically and crowed sensed data streams. Some of these data streams are structured in a way that serve only one predefined purpose and cannot be directly used for other means. Yet, there are emerging unstructured data such as context-based data from the internet and social media as well as credit card transactions that is not clear if they can be used to better understand the mobility patterns.

The analytical company Gartner states that by 2020 there will be over 26 billion interconnected devices. It is obvious, that they will produce massive amounts of meaningful data. Those data can be used for many applications such as real-time industrial equipment monitoring, traffic planning, automated maintenance, etc. Therefore, it is essential to develop modern system abstractions that allow us to resourcefully process huge and new data streams. This enormous amount of data urges the growth of integrated and insightful big and real-time data analytics Platforms.

The upcoming contemporary technology like digital twin, integrates historical data from past machine usage to the current data. It uses sensors to collect the real-time data, working status and other operational data attached to the physical model. These components send the relevant data via a cloud-based system to the other side of the bridge with the help of data analytics platform which produces the required insights. The big and real-time data analytics Platforms assist to perform useful operations on data analytics as a complete package. For this purpose, data analytics platform are used to acquire constructive insight from the huge volume of data.

Data analytics platform is an ecosystem of technologies and services that can help the businesses in increasing revenues, enhance operational efficiency, stabilize marketing campaigns and customer service efforts, respond more quickly to emerging market trends and gain a competitive edge over rivals. The data analytics platform finds the pattern and relationships in data by applying statistical techniques and communicates the results generated by analytical models to executives and end users to make decisions with the help of data visualization tools that display data on a single screen and can be updated in real time as new information becomes available. Big data and real-time data analytics platform supports the full spectrum of data types, protocols and integration to speed up and simplify the data wrangling process. The big data and real time platform provides accurate data, increase efficiency in the workspace, gives answers to complex questions along with security and hence it plays the key role in business analytics.

1. Introduction

Big Data generally refers the massive data sets where data is obtained from different sources such as sensor systems, telecommunications systems, experimental data, social network activities, surveillance camera, financial and business transactions [1]. For instance, Companies gather large amount of data about the customers, operations, suppliers and their transactions. Sensors located in various places such as vehicles, buildings, roads, and smart energy meters generate billions of bytes of real-time data. Moreover, billions of people across the globe render their contribution through social media platforms which eventually increases the availability and size of big data. It paves the great opportunity for using it efficiently in terms of achieving benefits in many real time applications. However, it causes various technological challenges in the aspect of organization, processing and analysis, storage capacity and management. With the availability of massive amount of information, the business organizations can understand the processing and operations of their businesses that helps in efficient decision making for improving the performance.

The available huge amount of big data can be used to improve the performance in various applications such as transportation, finance, healthcare, education, government and security purposes. Real time decision making is vital in order to increase the performance of different operations and services, profits in some special kinds of applications. For example, the applications such as military operation decision making, financial market trading and surveillance, intelligent transportation, emergency response, and smart grids need to perform rapid analysis for instantaneous decision making from the current and historical huge volume of data [2].

Untimely or slow data collection leads to delay in decision making significantly reduces the performance in real time big data applications. Such applications need to face lot of technical challenges in order to obtain optimal decision making in real-time processing.

In big data, the huge data sets are not able to be managed with typical database management systems because of its massive sizes. The size in big data ranges from few terabytes to many petabytes. Generally, big data applications searches for novel information and collects intelligence from the available data to achieve business benefits. Big data offers new different types of advances services that enrich the quality of life by reducing threats and serious risks. Big data can be categorized from the normal data with the following five characteristics: [3]

- Volume
- Velocity
- Variety
- Values
- Veracity

Volume: It was anticipated that 1 billion gigabytes of data are generated from various sources every day in 2012 and it is doubling in the count for every 40 months. The usage of smart phones, mobile devices, internet, etc. significantly contributes in gathering the massive amount of data. In another perspective, data is collected from different sensor systems that exist in factories, cars, buildings, roads and other environments.

Velocity: Immense amount of data are generated every second and that would be appended to already existing data sets. In many applications, the newly generated data will be included at the time of real time decision making. Organizing and manipulating the data which is gathered for making decision in real time applications is normally considered as a highly complex technical challenge.

Variety: The data that is usually gathered from different sources can be in structured or unstructured format of data. The data from various sources can be of images, messages, economic and political news, transactions, etc. The data that are relatively available are new because of the arrival of online social media and smartphones. Data is not generated uniformly and it may be periodically or frequently produced in a random manner even with different frequencies also.

Value: The vital features of the data are referred as a value that is worth of the data which are depend on processes they represent like probabilistic, random or stochastic. The value of data is not destroyed and can be reused by combining with some other data sets.

Veracity: The degree at which the data is precise, trusted and accurate is the veracity.

In this chapter, big data is analyzed from two different perspectives: Big data platform and real time applications of big data. Big data platform provides data governance, resource management and monitoring. Rest of this paper is structured as follows: Section 2 explains big data architecture and the components of big data architecture. Section 3 provides the details about the big data platform. Section 4 explains the conversion of big data to smart data to digital twin followed by real time applications in Section 5 and its challenges and real time response in Sections 6 and 7, respectively. The final conclusion is drawn in Section 8.

2. Big data architecture

From ancient India, Greece, Rome, Egypt and America, the architecture occupies the vital part of any era all around the world. Here, the big data architecture is discussed that will hopefully leaves its impression in the upcoming millennia. Big data architecture overcomes the limitations of traditional database systems and it is designed to handle the input data, its processing and analysis.

In recent years, the landscape of data has changed. The method of collecting data is varying as several data is collected at a rapid pace while some data arrives in bulk in the form of historical data. In addition to traditional database workload like Batch processing of big data sources at rest and real-time processing of big data in motion, big data involves predictive analytics, machine learning and interactive exploration of big data.

2.1 Components of big data architecture [4]

The big data architecture accommodates some or all the following components (Fig. 1).

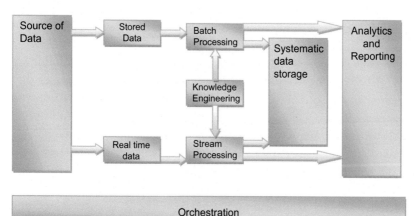

Fig. 1 Components of big data architecture.

Source of data is the basic of big data architecture. The source of data can be relational databases, web server log files or a real-time data sources.

Data for batch processing operation are stored in distributed database that accommodates huge volume of data of different formats, often called a *data lake*. Example is Microsoft Azure Data Lake Store [5].

Big data operation deals long-running batch jobs and it involves scanning source files, process and giving the output. Some of the batch processing big data tools is Hadoop, Pig, Azure Data Lake Analytics, Hive, Java, Scala and Python programs.

If big data solutions include real-time sources, the architecture should include storage facility for real-time messages for stream processing. The tools include Azure IoT Hub, Azure Event Hubs and Kafka.

- *Stream processing.* Data are gathered from different sources and subjected to pre-processing by means of actions like filtering, aggregation, regression, clustering, etc. After pre-processing, the data are written to an output sink. A managed stream processing is provided by Azure Stream Analytics to handle perpetually running SQL queries which can operate on unbounded streams. Apart from Azure Stream Analytics, open source Apache streaming technologies such as Storm and Spark Streaming in an HDInsight cluster can also be used.

- *Analytical data store.* After pre-processing, the data is stored in structured format and it can be analyzed using analytical tools. Kimball-style relational data warehouse is used to store the analytical data. Apart from this data warehouse, the low-latency NoSQL technology HBase or Hive database can be used. Azure SQL Data Warehouse is used for cloud-based large-scale applications. Interactive Hive, Spark SQL and HBase are HDInsight used to serve data for analysis.

  Analysis and reporting. The main objective of the big data solution is to visualize output. Reporting is the outcome of data analysis which gives efficient solutions. To empower this multidimensional OLAP cube that acts as data modeling layer is introduced in the big data architecture. It also supports data visualization tools like Tableau, Qlikview or Power BI. Data scientists explore the data from analysis and reporting tool and hence some service providers like Azure supports open source software like Jupyter which is an interactive computing platform. Some large-scale data exploration utilizes servers like R as a standalone or in combination with spark.

- Orchestration. As big data deals with parallel processing of data it need a meta scheduler or an orchestrator to organize the activities. Orchestrator navigates various applications by governing error management and make easy interaction with other solutions. Orchestrator enables secure transferring of data through firewalls and balances the overall efficiency of the load. Sqoop, Azure Data Factory and Apache Oozie are some of the orchestration technologies are used currently.

3. Real-time big data analytics platforms

Big data platform is a combination of technologies and services that analyze lively and complex data. Hence choosing the correct software and hardware technologies is a crucial part [6]. The necessity of big day has evolved many platforms recently including Hadoop, Kira, Kappa, Lambda, Big Data Europe (BDE) [7], Cloudera and Hortonworks.

3.1 Classification of platforms

The platforms for big data is classified into three types, namely, batch processing, real time processing and interactive analytics. Batch processing platform is the most deployed big data platform that requires more time to process data as it carries out complex computations. Real time processing needs rapid data processing and streaming is required in solutions that need least latency. Users can remotely access dataset in interactive analytics.

In other way, big data platforms can be classified along two dimensions as shown in Fig. 2:

1. Based upon the throughput of data processed per unit time and
2. Latency time to process a unit of data.

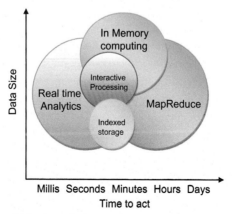

Fig. 2 Classification of big data platform.

Hadoop is a volume-driven platform which can be scaled up to terabytes of data and are able to process 100 MB/s of data per second. Since Hadoop is a batch processing platform it takes time to generate results.

3.1.1 Hadoop

Apache Hadoop is an open source framework implementation of MapReduce that do distributed processing of huge volume of data using simple programming models. Google designed MapReduce scale up from single servers to many machines and offers local computation and storage. MapReduce concept is inspired by map and reduce primitives of Lisp Language.[8]. In Hadoop, a problem or query is split by Hadoop recursively into small chunks to be solved. These small units are distributed on system nodes for execution through the coordination of a master node and working nodes [9].

In Hadoop, the libraries are constructed in a way to detect and handle failures at the application layer. Hadoop modules include HDFS, YARN and MapReduce. Hadoop also has abstractions that operate in its ecosystem; they are HBase, Chukwa, Flume, Hive, Zookeeper and Sqoop. These Hadoop abstractions and modules functions on big data value chain ranging from acquiring data, storage, processing, analysis and management. Companies require Hadoop due to its cost effectiveness, scalability, fault tolerance and flexibility. MapReduce is a version of directed acyclic graph (DAG) organized as two functions [10].

The primary operation is a map function which divides an element into number of key value pairs. The secondary operation is the reduction function which merges the values into a consolidate form (Fig. 3).

3.1.2 Lambda architecture

Lambda architecture contributed by Nathan Marz is one of the best architecture in real-time data processing intended to handle low-latency reads and updates in a linearly scalable and fault-tolerant way. The software design pattern combines online and batch processing in a single framework. This pattern is well suitable for the applications which have time delay in data collection. The batch processing is followed to search for the behavioral patterns from the data sets as per the user requirement. The lambda architecture consists of three layers:

(1) Batch processing is significant for pre-computing massive amount of data sets.

(2) Real time computing minimizes latency by calculating real time data rapidly.

(3) The layer responds queries and obtains the result.

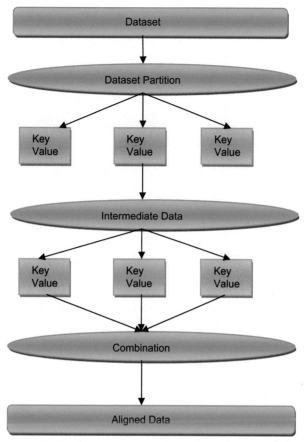

Fig. 3 Hadoop architecture.

Lambda architecture assists in reducing the cost of data processing by triggering the specific part of the data which need batch or online processing. The available dataset is partitioned that allows different kinds of scripts to be executed. The lambda architecture is best suited for big data processing for analyzing the data gathered from sensors. The online stream is followed in order to identify the data anomalies before processing it and to validate the data whether it is accurate or not. Various data patterns can be studied from the data stored in the database over a time period. The cost is reduced by breaking down the problem on larger data sets into manageable units. This architecture is much efficient in gathering and analyzing sensor data to obtain the solution by processing massive data sets. The following figure shows the Lambda architecture (Fig. 4).

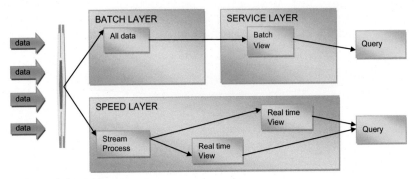

Fig. 4 Lambda architecture.

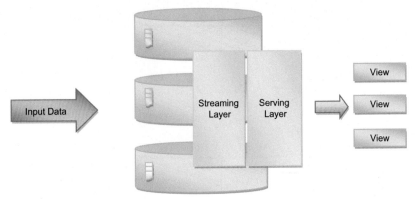

Fig. 5 Kappa architecture.

3.1.3 Kappa architecture

Kappa architecture is a simplified version of lambda architecture developed by Jay Kreps in 2014 [11]. Kappa architecture resembles lambda architecture without batch processing system.

Kappa architecture is not a substitute version of Lambda architecture but in contrast it is considered as an alternative which is used in the situation where batch processing layer is not needed to meet standard quality of service. To reinstate batch processing, the incoming data are passed through the streaming system hastily. The data are stored in an append-only permanent log from where the data is streamed to a computational system and forwarded into auxiliary stores for serving.

Kappa architecture employs single code path for the two layers which reduces system complexity [12] (Fig. 5).

The above figure shows the Kappa architecture and it is composed of stream processing layer and serving layer which is used to query the results.

Kappa architecture is used in the situation when multiple queries. Kappa architecture is employed for the data processing enterprise models where multiple queries are logged in a queue to be catered against distributed file system storage. The order of the query is not pre-programmed. The database can be accessed by the stream processing platform at any time. The Kappa architecture is implemented using technologies like, HBase, Samza, Apache Storm, Kafka and Spark. Kappa architecture provides processing and data consistency as it provides a reliable and real-time execution of its log system.

3.1.4 Kira architecture

Improving the computational performance and minimizing I/O cost are significantly focused while designing KIRA toolkit due to the flexibility in code reusability. A process in KIRA can be a Spark Worker, Spark Driver, or a HDFS daemon. KIRA supports both single as well as multiple drivers and that is executed on the top of Spark. The input files are available in the file system and the SEP library exists in all worker nodes. For executing KIRA, the parameters, library dependencies and the compiled program are submitted to the Spark Driver. It manages the task scheduling, data flow, and control flow by managing the coordination among the Spark workers. The parallel/distributed file system is accessed by the Spark Driver for managing I/O operations and metadata that are disseminated in parallel among Spark Worker Nodes. If a particular task is executed and the computation is performed by the workers by invoking SEP library.

Computation: There are three approaches mainly considered while implementing the Source Extractor algorithm in KIRA. They are:

i. Re-implementing the algorithm from the beginning.
ii. Combining programs as a single unit without any modification.
iii. Re-structure the C-based SExtractor implementation.

If the functionality of CExtractor is re-implemented using Apache Spark's Scala API, SExtractor can be executed in parallel where the computational efficiency is reduced. The original executable code cannot be modified in the monolithic approach, but it can be integrated with the original code at this level. For instance, astronomers can enhance the accuracy level by executing multiple iterations of the source extraction. However, the hardcoded logic permits the user to execute the source extraction once. An alternative library based model is approached in order to avoid the limitations of control-flow flexibility and the legacy code base is reused in this approach (Fig. 6).

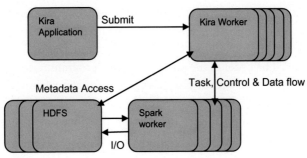

Fig. 6 Kira architecture.

4. Big data to smart data to digital twin

Industry 4.0 and Industrial Internet of Things (IIoT) is becoming exhortation focuses greatly on machine learning, automation, interconnectivity and real-time data. Industrial Internet of Things (IIoT) combines production technology with smart digital technology, big data and machine learning to create a novel ecosystem for companies. Digital twin is the digital representation of physical object which plays a major role in Industry 4.0 [13]. The concept of digital twin was first introduced by Grieves in 2003 at University of Michigan [14].

The concept of digital twin was first presented by Grieves at one of his presentations about PLM in 2003 at University of Michigan. The digital twin accommodates three components, namely, physical entities, virtual models and the data which connects physical and virtual models. Big data plays a vital role in the digital twin technology. The convergence of big data and digital twin is breaking the barriers among various phases of product life-cycle while minimizing the product development and verification cycle. The data in the digital twin technology is analyzed either in real time or in batch or as a combination of both. This data handling concept resembles lambda architecture and it is well suited to process data in digital twin technology. Batch analytics is a method of processing huge volume of data gathered over a long period of time. For instance, batch analytics for digital twin provides a visualization of minimum and maximum temperatures for a physical asset collected over a period of time. Real-time analytics might be used to alert a central command center about physical damages incurred to an asset. Hence big data and digital twin can be combined to endorse smart manufacturing (Fig. 7).

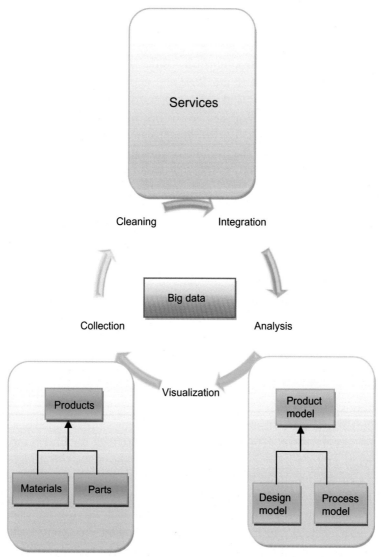

Fig. 7 Union of big data and digital twin in smart manufacturing.

Digital twins are implemented using Lambda architecture [15] because of its fault tolerance nature, data ingestion characteristics, deployability and adaptability. In digital twin process, the field system is collecting data from sensor based IOT devices. The collected are filtered and synthesized with the help of IOT broker and stored inside a NoSQL DBMS.

5. Real time applications

Real time applications normally differ from regular applications because of one significant attribute. Real time applications are highly dependent on instantaneous input and rapid analysis to make a decision in a very short time span. Moreover, if decision is not mode at right time then it becomes useless. At last, it is mandatory to produce all necessary data for efficient decision making and that is done in a reliable way. Some of the real-time big data analytics applications are discussed.

5.1 Intelligent transport system

Nowadays, different sensing technologies exist to track the traffic conditions in big crowded cities. In intelligent transportation systems, the sensor types are mainly categorized into vehicle sensors and road sensors. Examples of vehicle sensors are proximity sensors, GPS systems, speedometers and on-board cameras [16]. Road sensors include vehicle inductive loop sensors, road tube axle sensors, road monitoring cameras, piezoelectric axle sensors and capacitance mats [17]. Such kind of sensors are interlinked with communication technologies such as WiFi, GSM, Bluetooth and Satellite communications to monitor the different environmental conditions such as location of the vehicle, driving attitude of drivers, average speed, road conditions, etc.

The data generated from vehicular and road sensors can be used to afford transportation services in big smarter cities. One instance for kind of services is providing information about shortest path from the current location to the destination depending on the present traffic conditions. This type of service is not able to provide without having the complete knowledge about the traffic conditions on the roads [18]. Moreover, such services are essential for regular vehicles and it is extremely significant for emergency vehicles such as ambulances, fire engines, police cars, etc. The present GPS system provides the best path availability to the drivers, however, due to sudden changes such as roadblocks or accidents it may not respond quickly. In addition, on-time information is not reached to clear the road to offer the way for emergency vehicles.

In advanced intelligent transportation services, end users are informed about efficient routes from the present location to any destination point. The eminent energy efficient routes can be determined by not only observing the traffic conditions, in addition few characteristics such as energy consumption in the vehicles, driving behavior of the driver, and previous traffic experiences are to be considered as significant factors. The real time dynamic solution is mandatory in traveling salesman problem to define the routes of delivery points in such a way that the delivery time is reduced. The route can be dynamically updated depending on the present traffic conditions during the travel in order to minimize the total delivery time.

5.2 Financial market trading and surveillance

The immense amount of financial data are created for trading from various sources such as currency exchange rates, commodity prices, interest rates and multiple markets. The financial organizations can utilize the dynamically generated big data to identify threats and make efficient decision in order to avoid any serious impact. For instance, anticipating increase or decrease in price before the change actually happens. The situation can be handled securely by taking the timely action before the prices drop. The sold-out securities are re-purchased at low price and it could be focused to increase the profit. An efficient forecast models are required to manage these opportunities and this model relies on both historical and current information.

On-time detection is mandatory to identify illegal and fraudulent activities such as price rigging, uninformed insider trading and market manipulation aids in preventing such activities rapidly thus it secures investors transactions and improves market performance. The efficient detection process is vital in order to prevent threats and well utilization of financial opportunities. Such things can be discovered by humans, whereas it requires long time, less opportunities and threats would start to have an impact before it could be prevented. Recently, few automated trading system are developed to deal with financial opportunities. Some real time fraud detection systems are able to detect threats in a timely fashion. The volume of financial data grows massively and changes dynamically, that faces various technical challenges.

5.3 Military decision making

The process in military domain is too complex and dynamic. During war, it is important to gather correct information at the right time about the current situation to make quick decisions [19]. In wars, number of different equipments, communication systems and vehicles are used. Military vehicles can be armored vehicles, tanks, manned and unmanned aircrafts, underwater vehicles, military boats and ships, transportation and logistics vehicles.

The vehicles, equipment, soldiers and other resources are in different locations. Information about the opposite side also needs to be gathered periodically in order to take responsive actions against attacks. The location of such kind of resources can be tracked in real time through satellites, GPS technologies, aircraft remote sensors, etc.

5.4 Crowd control

It is the requirement of emergency response team when the largest events take place. For instance, events such as major concerts, sports games, outdoor celebrations, parades, etc. Predicting crowd movements and providing quick decisions such as facilitating some streets for pedestrian movement, arranging more parking lots, increasing the number of police count in some specific areas becomes significant. This can be accomplished using tracking and sensing technologies. Traffic management using traffic sensors and vehicle monitoring permits for gathering information about the incoming traffic thus a better control can be afforded in the heavy traffic area where it is directed. Nowadays, GPS and location tracking applications are used for monitoring the people and their movement.

The tracking applications in real time provide an accurate view of everyone in the overcrowded areas. Police forces can be distributed in a better manner to have an optimal control and to provide the services in the best possible locations. Moreover, the probabilities for catastrophic incidents and accidents are more in such kind of areas. Real time analysis plays a key role in locating the spot of an accident and response strategies are activated immediately such as alerting the people in the area to get rid of the place, clearing paths for response vehicles and the support can be provided in a timely manner.

Emergency response in a large-scale: in disasters such as floods, volcanoes, earthquakes, wars, terrorist attacks, immediate correct actions should be taken within a short span to help those who were affected. In such crucial scenarios large amount of information will be sent to a control center which is gathered from normal people and other officials. Some information will also be obtained from the deployed sensors, satellites, UAVs and robots. By utilizing the available equipment and human resources, emergency control center will take steps to relieve the injured as soon as possible. Equipment resources include temporary communication devices and emergency vehicles such as ambulances, fire engines, boats, UAVs, etc. The real time decision support system gathers information and processes to facilitate an optimal resource allocation and management.

Prior warning about natural disasters: it clearly says that prior warning about disasters will definitely save the lives of many people. The warning systems include time critical processing of data collected from underwater sensors, ground sensors, remote sensors those installed in satellites and aircraft vehicles uses geographical data, weather information to predict when and where the disaster may happen. Indian Tsunami Warning system includes ocean based sensors to identify tsunamis, sensors to detect earthquake, satellites to declare weather information, and geographic maps to detect the place where the tsunami will hit.

5.5 Smart grid

It is an electrical grid system utilizes technology to gather information about consumers and suppliers behavior. It improves reliability, sustainability, efficiency of the electric power generation and distribution by receiving feedback with two-way communication technology among producers and consumers. Smart sensors and meters are placed on transmission, production, and distributed systems to obtain the granular real time data about the power consumption and faults.

6. Challenges in real time big data applications

Real time applications need to initiate quick response actions within a specific time frame in the targeted domain. Unlike regular applications, real time big data applications follows closed-loop approach for implementation and the processing is based on previous and current situations. To understand the design, implementation and operation of real time big data applications, the primary steps to be performed as mentioned below:

- Event transfer
- Situation discovery
- Real-time analytics
- Decision making
- Real-time responses

6.1 Event transfer

The events in all distributed applications are transferred as raw or aggregated or filtered events. Such events are transmitted to intermediate or centralized point for further processing and aggregation before transferring to the centralized decision making unit. The centralized approach is preferable if no huge number of recently generated events and no limitations on the availability of network resources. The distributed approach is suitable to transfer the generated events to a single point location. In both open and closed loop approach, event aggregation and filtering are essential to minimize the processing time and network traffic without affecting the accuracy as well as optimality of decision making process in real time big data applications.

6.2 Situation discovery

It is focused to identify the operational situation of the current events along with its exceptions in real time business. For example, sudden traffic in some specific area in smart intelligent transportation application, a drastic drop in the stock price of a certain company in financial trading application, unexpected huge power consumption in a specific area in smart grid application. The detection process is based on some predefined rules set by the developers in real time big data applications. Decision making or analytical modules can modify, add or delete some rules in order to alter the operation of situation discovery. In the context, the situation discovery process can modify the policies for filtering and aggregating the event transfer.

6.3 Real time analytics

This step is significant to find out the root causes for operational and exceptional situations in real time analytical services. The analytical process may have single or integrated services that predict the performance of the business environment and risks are assessed before changing the operation. Dealing with such huge data set is technically challenging and if it viewed as smaller data sets in an abstract form significantly reduces the execution time of processing.

6.4 Decision making

It is highly important to choose the best option for increasing the profit in order to increase the present business situation and the most appropriate action as a response is vital in the business environment. The operational rules are defined by the domain experts and inferred from the strategic decisions for efficient decision making. Intelligent and interactive response aids in evolving the business situations in a best way. The main challenging part lies in defining operational and business rules which facilitate timely and efficient decision making.

6.5 Real time responses

It comprises of initiation, execution and tracking the activities occurred in real time decision making process. For instance, to purchase a specific unit of stock quantity for a company, it involves the pro Big Data generally refers the massive data sets where data is obtained from different sources such as sensor systems, telecommunications systems, experimental data, social network activities, surveillance camera, financial and business transactions. For instance, Companies gather large amount of data about the customers, operations, suppliers and their transactions. Sensors located in various places such as vehicles, buildings, roads, smart energy meters generates billions of bytes of real-time data. Moreover, billions of people across the globe render their contribution through social media platforms which eventually increases the availability and size of big data. It paves the great opportunity for using it efficiently in terms of achieving benefits in many real time applications. However, it causes various technological challenges in the aspect of organization, processing and analysis, storage capacity and management. With the availability of massive amount of information, the business organizations can understand the processing and operations of their businesses that helps in efficient decision making for improving the performance.

The available huge amount of big data can be used to improve the performance in various applications such as transportation, finance, healthcare, education, government and security purposes. Real time decision making is vital in order to increase the performance of different operations and services, profits in some special kinds of applications. For example, the applications such as military operation decision making, financial market trading and surveillance, intelligent transportation, emergency response, and smart grids need to perform rapid analysis for instantaneous decision making from the current and historical huge volume of data. Untimely or slow data collection leads to delay in decision making significantly reduces the performance in real time big data applications. Such applications need to face lot of technical challenges in order to obtain optimal decision making in real-time processing.

7. Conclusions

Big data has the ability to analyze and store huge volume of data for various applications. To analyze the massive data, a platform is necessary to handle data from multiple distributed sources. Big data platform is an

Information Technology solution that holds the features and capabilities of various big data applications within a single solution. In this chapter the architecture of big data is discussed as it need special attention because it handles massive volume of data and hence a robust architecture is necessary. The platforms are selected based on the method of processing of data. The platforms for batch processing, real time and combined scenario are discussed elaborately. The role of big data in shaping digital twin technology with Lambda architecture and the reason to why this platform is preferred is neatly explained. The real time application of big data in various field and its challenges are discussed here.

References

[1] Galar D, Kumar U: *EMaintenance,* 2017, Academic Press.
[2] Borko F, Flavio V: Introduction to big data. In Borko F, Flavio V, editors: *Big Data Technologies and Applications,* 2016, Springer International Publishing, pp 3–11.
[3] Alonso-Betanzos A, Gámez J, Herrera F, Puerta J, Riquelme J: Volume, variety and velocity in data science, *Knowl Based Syst* 117:1–2, 2017.
[4] Docs.microsoft.com: Big Data Architectures, 2019: [online] Available at https://docs.microsoft.com/en-us/azure/architecture/data-guide/big-data/.
[5] Docs.microsoft.com: Big Data Architecture style—Azure Application Architecture Guide, 2019: [Online]. Available, https://docs.microsoft.com/en-us/azure/architecture/guide/architecture-styles/big-data.
[6] Singh D, Reddy CK: A survey on platforms for big data analytics, *J Big Data* 2(1):8, 2015.
[7] Jabeen H, Archer P, Scerri S, Versteden A, Ermilov I, Mouchakis G, Lehmann J, Auer S: Big data Europe. In Proceedings of the Workshops of the EDBT/ICDT 2017 Joint Conference.
[8] Dean J, Ghemawat S: MapReduce, *Commun ACM* 51(1):107–113, 2008.
[9] Chardonnens T, Cudre-Mauroux P, Grund M, Perroud B: Big data analytics on high velocity streams: a case study. In *2013 IEEE International Conference on Big Data.* 2013, pp. 784–787.
[10] Estrada T, Zhang B, Cicotti P, Armen RS, Taufer M: A scalable and accurate method for classifying protein–ligand binding geometries using a MapReduce approach, *Comput Biol Med* 42:758–771, 2012.
[11] Kappa Architecture, Kappa Architecture—Where Every Thing Is A Stream, *Milinda. pathirage.org,* 2019. [Online]. Available: http://milinda.pathirage.org/kappa-architecture.com/.
[12] Kreps J: *Questioning the Lambda Architecture,* 2014, O'Reilly, pp. 1–10.
[13] Hochhalter J, Leser WP, Newman JA, Gupta VK, Yamakov V, Cornell SR, Willard SA, Heber G: *Coupling Damage-Sensing Particles to the Digital Twin Concept, 2014:* 2014, NASA Center for AeroSpace Information Available, https://ntrs.nasa.gov/search.jsp?R=20140006408.
[14] Grieves M: Digital Twin: Manufacturing Excellence Through Virtual Factory Replication, 2014: White paper. Available, http://www.apriso.com.
[15] Kiran M, Murphy P, Monga I, Dugan J: Baveja, Lambda architecture for cost-effective batch and speed big data processing, *2015 IEEE International Conference on Big Data (Big Data)* 2785–2792, 2015.

[16] Tyburski RM: A review of road sensor technology for monitoring vehicle traffic, *ITE J* 59(8):27–29, 1988.

[17] Ferreira M, Fernandes R, Conceição H, Gomes P, d'Orey PM, Moreira-Matias L, Gama J, Lima F, Damas L: Vehicular sensing: emergence of a massive urban scanner. In *Sensor Systems and Software*, Berlin Heidelberg, 2012, Springer, pp 1–14.

[18] Biem A, Bouillet E, Feng H, Ranganathan A, Riabov A, Verscheure O, Koutsopoulos H, Moran C: IBM infoSphere streams for scalable, real-time, intelligent transportation services. In *Proceedings of the 2010 ACM SIGMOD International Conference on Management of Data*, 2010, ACM, pp 1093–1104.

[19] Feith DJ: *War and Decision*, 2009, HarperCollins.

About the authors

Ms. Indrakumari Ranganathan is working as an Assistant Professor, School of Computing Science and Engineering, Galgotias University, NCR Delhi, India. She has completed M.Tech in Computer and Information Technology from Manonmaniam Sundaranar University, Tirunelveli. Her main thrust areas are Big Data, Internet of Things, Data Mining, Datawarehousing and its visualization tools like Tableau, Qlikview.

Dr. Poongodi Thangamuthu is working as an Associate Professor, in School of Computing Science and Engineering, Galgotias University, NCR Delhi, India. She has completed Ph.D in Information Technology (Information and Communication Engineering) from Anna University, Tamil Nadu, India. Her main thrust research areas are Big Data, Internet of Things, Ad-hoc networks, Network Security and Cloud computing. She is pioneer researcher in the areas of Bigdata, Wireless network, Internet of Things and has published more than 25 papers in various international journals. She has presented paper in National/International Conferences, published book chapters in CRC Press, IGI global, Springer, and edited books.

Dr. Suresh Palanimuthu received B.E. degree in Mechanical Engineering from University of Madras, India in 2000. Subsequently received his M.Tech, and Ph.D., degrees from Bharathiar University, Coimbatore in 2001 and Anna University, Chennai in 2014. He has published about 25 papers in international conferences and journals. He is a member of IAENG International Association of Engineers. He is currently working as Professor, Galgotias University, Uttar Pradesh, India.

Dr. Balamurugan Balusamy Completed Ph.D at VIT University, Vellore and currently working as a Professor in Galgotias University, Greater Noida, Uttar Pradesh. He has 15 years of teaching experience in the field of computer science. His area of interest lies in the field of Internet of Things, Big data, Networking. He has published more than 100 international journals papers and contributed book chapters.

CHAPTER EIGHT

Air pollution control model using machine learning and IoT techniques

Chetan Shetty, B.J. Sowmya, S. Seema, K.G. Srinivasa
M S Ramaiah institute of Technology, Bengaluru, Karnataka

Contents

Abstract

Problem with the automobile engines continues to increase in a very large scale. Every vehicle has its own emission but problem arises when the emission occurs beyond standard values. Even though lot of changes has been made in the consumption of fuel, increasing urbanization and industrialization contribute for the poor air quality. With the technical advancements in machine learning, it's been possible to build predictive models for monitoring and controlling pollution based on the real-time data. With this, we are using IoT techniques for monitoring the emission rates of vehicles. A predictive model is built on the real-time data available, predicting the values of carbon monoxide. Sensors are embedded in the vehicles to measure the pollutants levels. By using the monitoring techniques, vehicle details such as location, owner is notified with the current situation of pollution in his location and his vehicle emission rate contributing to environment. Machine learning model is used for the prediction of pollution level in the vehicle location based on the previous data and the current data obtained by the sensors. Here the pollutants level can be controlled using smart emission surveillance system. The system shoots beyond threshold value taken from the Bharat Stage emission standards then automatically a notification will be sent to the vehicle owner. The emitted level will be monitored and the fuel supply to the engine will be cut off using solenoid valve at the same time.

1. Introduction

Air pollution is one of the gravest problems faced by the environment especially by densely populated countries like India. The rapid growth of population has led to an increased usage of vehicles. These vehicles use fuels which undergoes incomplete combustion to emit toxins. The principal emissions from motor vehicles contains harmful gases which contributes to greenhouse effect.

Automotive emissions include CO_2, carbon monoxide (CO), hydrocarbons (HC), nitrogen oxides. The magnitude of each of these emissions relies on the mixture of vehicle/fuel, weather conditions, and driving patterns. Under emission standards, the emission limits for a specific car differ depending on the vehicle's weight, its type of gas, and whether it is a passenger or a vehicle carrier. Diesel cars have a greater nitrogen oxide emission limit, while petrol cars have a greater carbon monoxide emission limit.

The main idea is to measure the emission levels from the motor vehicles and control the pollution. Sensors are used to detect the gases released from the exhaust. The levels of the different gases are measured and compared with the standard values. If the amount of pollutants is higher than the threshold level then the LED is turned on automatically, which is controlled by Arduino and vehicle owner will be sent a notification. The fuel supply is stopped using fuel injector making the vehicle immobile.

The purpose of this work is to build a robust system that can keep check on the amount of pollutants released by the vehicles into the atmosphere. Sudden environmental changes in pollution is one of the major contribution to environmental pollution. Increased number and usage of vehicles has reduced the quality of air and its surroundings and has even led to catastrophic problems in human beings.

Hence, a technical solution needs to be developed to reduce air pollution from automobiles.

Problems include:
- How to find out the amount of pollutants released by vehicle?
- How to reduce the emission of toxic gases from vehicles?

- How to assist the owner in keeping a check of his contribution to air pollution?
- How to make the system affordable to all?

These above-mentioned problems can be solved provided we know how to deploy technology.

- The solution must be cost effective and flexible to the user.
- It must consume less power supply.
- The solution must be eco-friendly.
- The solution must be durable and fault tolerant.
- Since the solution demands lots of hardware component, it must be compact and user friendly.

1.1 Objectives of the project

The main objective of the project is to reduce air pollution. The project aims at enabling facilities such as:

- A device that collects the value of pollutants emitted by the vehicle

Multiple sensors are placed together to form a device which collects the data about all the different pollutants emitted by the vehicle. The device is placed at the exhaust of a vehicle to ensure accurate readings.

- Notifying the owner of the vehicle about excessive pollutants being emitted by their vehicle

Whenever the emission level of pollutants crosses the standard levels set by NAQI there is a message sent to the owner of the vehicle. The message notifies the owner about the current status of the environment and the impact of excessive emission. The aim of message is to intimate the owner that their vehicle needs servicing.

- A device that stop the flow of fuel from the fuel tank to the engine

After the owner has been notified about the excessive pollutants emitted from their vehicles. There is a waiting phase to ensure that the owner take appropriate measures to control the emission. If there are no changes made and the emission continues to be greater than the standard values we use a solenoid to block the flow of fuel.

The system uses sensors to detect the gases released from the exhaust of vehicle. Depending on the threshold values set, led glows depending on the amount of gases released into the atmosphere. If the quantity released is beyond the threshold value, green led glows else the red.

The values of toxic gases released is noted and sent as text message using GSM board to the owner of the vehicle. If any of the value of gas exceeds threshold value, the fuel supply is cut off using the solenoid valve preventing the vehicle from being used until it's given for service. The current scope of the project is that we can retrieve a better processed data from sensors and send the processed data to server for maintenance by refactoring the code. Hardware compaction is the major current scope of the project. The other possible scope of the project is real-time data analysis which leads to better efficiency. An Android application for the routing of vehicles through less polluted areas is also one of the scopes.

2. Literature survey

India is the world's second-largest country in terms of population. Transportation sector is a key component in its rapidly growing economy. This project aims at preventing and predicting the air pollution produced by automobiles using Arduino board, three gas sensors, namely, MQ-2, MQ-7 and MQ-135, GSM module and solenoid valve. It is mainly an IoT based project. The sensors sense the gaseous levels of emissions released from the exhaust of the vehicle. Based on the threshold value, the respective led glows for each sensor depending on the quantity of toxic gases which it can sense. An alert message is sent to the owner warning him that his vehicle has exceeded the safe emission standards of BS IV. We propose this system as this idea is analyzed and conceptualized by referring a few IEEE papers and by applying our own ideas on prevention of air pollution and controlling it. This system has a major feature which aids in curbing the air pollution to a huge extent. The system cuts/chokes down the fuel supply from the fuel injector using the solenoid valve when the emission rate is higher than the given threshold value. IoT is an emerging field and its technology assists in automating almost everything. Hence, we use the benefits of this field in controlling a major problem faced by the environment that is air pollution. This system has a great possibility of bringing a revolutionary change in the concept of prevention and control measures for air pollution.

M.U. Ghewari et al. [1] discusses about how to monitor air pollution on roads and track vehicles which cause pollution over a specified limit. The great amount of particulate and toxic gases is produced due to increase in industrialization and urbanization. There is poor control on emissions and little use of catalytic converters. The serious problem that has been around for a very long time is increase in usage of automobiles. Internet of things (IoT) [2] is being used to address this problem. Here, combination of electrochemical toxic gas sensors, wireless sensor networks and radio frequency identification (RFID) [3] tagging system are used to monitor car pollution records anytime and anywhere. Few locations to be monitored with usually high volume of traffic are identified. The RFID readers are being placed on the either side of a road for each location with a fixed distance between them. A passive RFID tag is equipped in each vehicle passing through road. Sensor nodes, composed of gas sensors, are placed on the roadside. The sensor nodes are identified and are addressed by unique IP address [4]. These sensors gather data continuously and sent to the server wirelessly. Whenever the sensors sense sudden rise in pollution, search is initiated for corresponding RFID tags, i.e., vehicles which are causing pollution are identified using the tag attached on them. These RFID readers detects a car passing by it. The RFID readers identify specific tag number and transmit the same via the GPRS [5] the server. This system also generates an alert when the pollution level increases. Then, the appropriate actions are taken by the authorities accordingly. The authorities monitor and analyze all the gathered data. The proposed framework is as shown below (Fig. 1).

A graph for pollutant data for various vehicles is shown below. Once the level of the pollution exceeds permissible level, motorists may be advised to avoid that particular area. It may be done using the same Internet of things. It may enable to reduce the pollution level over a certain span of time. This framework may be integrated as an enabling tool to design intelligent transportation system for smart city [6] (Fig. 2).

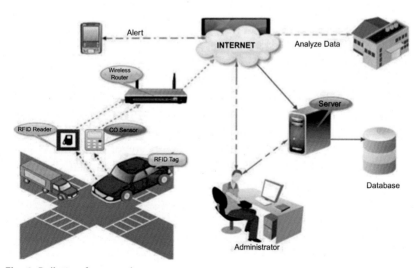

Fig. 1 Pollution framework.

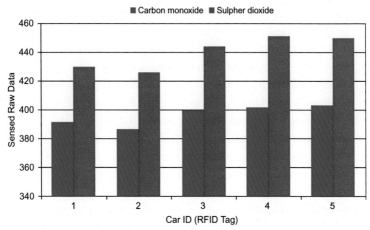

Fig. 2 Pollution level monitoring.

B. Hunshal et al. [7] discuss the various sensing criteria, test processes, environmental parameters and microsystem-based realization needs. Gas sensors have become an essential element of internal combustion engine control systems to provide data for air-to-fuel (A/F) feedback control to improve car efficiency and fuel economy as well as lower emission concentrations. Increasingly strict restrictions on evaporative emissions and on-board diagnostic requirements [8] (OBD), including catalyst monitoring, require monitoring of exhaust gas constituents [i.e., carbon monoxide (CO), hydrocarbons (HCs), and nitrogen oxides (NOx)]. MOTOR cars have two primary emission kinds: (1) tailpipe exhaust and (2) fuel system evaporative emissions. The only exhaust gas sensor that is currently used on cars is the oxygen sensor that has been used widely in cars for over 15 years. The primary function of this sensor is for feedback control of the air-to-fuel ratio [9] (A/F) to maintain the gasoline/air mixture close to stoichiometry in order to minimize emissions. The tailpipe emissions are present only when the vehicle is operated (as opposed to the evaporative emissions) and the undesirable emissions can be divided into four main categories: (1) well over a hundred different species of HCs (including oxygenated "HC," such as aldehydes and ketones); (2) carbon monoxide (CO); (3) oxides of nitrogen (NO); (4) particulates (mainly from diesel engines).

For low tailpipe emissions, the precision of the air-to-fuel feedback control system (under static and dynamic driving circumstances) is of paramount significance as the catalyst's effectiveness is significantly peaked as a function of A/F. One idea [10] is to add an extra feedback loop to the air-to-fuel control system to further optimize the A/F for lowest emissions. This can be achieved by measuring behind the catalytic converter HCs (and/or CO) and NO. To obtain the smallest possible emissions, the air-to-fuel feedback scheme would then minimize the (normalized) difference between the two sensor signals. This works because the level of combustible gas concentrations substantially increases under fuel "rich" conditions, while substantially increased levels of NO indicate fuel "lean" conditions. By constantly measuring the exhaust gas concentrations as well as other readily available engine operating parameters, such as temperature, speed and load, and mass air flow, an actual mass of emissions can be obtained [11]. This information could be used to monitor the tailpipe emissions of the entire combustion system.

S.P. Bangal et al. [12] use a vehicle identification and detection strategy based on unintended electromagnetic emissions. When operating, cars with inner combustion motors radiate the vehicle's distinctive electromagnetic emissions. Emissions rely on electronics, harness cables, type of body, and many other characteristics. Since each vehicle's emissions are unique, they can be used for purpose of identification. This article explores a procedure based on their RF emissions to detect and identify cars. Measured emission information collected parameters such as the average magnitude or normal magnitude deviation within a frequency band. These parameters have been used as inputs to an artificial neural network [13] (ANN) trained to define the car producing the emissions. The approach was tested with emissions from a Toyota Tundra, a GM Cadillac, a Ford Windstar, and ambient noise captured. When using emissions capturing an ignition spark event, the ANN was able to classify the source of signals with 99% accuracy. Using neural networks, identification was allowed. In order to highlight distinctions between cars and ambient noise, several parameters were obtained from the measured emission spectrograms. The most significant parameters were the standard deviation and amount of pulses in a frequency band. A 99.3% identification rate could be accomplished using these two parameters alone when a spark event was captured. When a spark event was not captured, however, the neural network was unable to successfully identify the responsible vehicle. It is possible that detection of vehicles without using the ignition pulse [14] could be accomplished if "noise-free" measurements of the vehicle were available to better train the network and to help form more useful parameters that characterize the vehicles in this case. S. Smurtie et al. [15] presents a car emission surveillance scheme based on IoT. Due to cars, the primary source of atmospheric taint occurs.

Using empirical scrutiny, the ritual mechanized air monitoring scheme is highly rigorous, but inexpensive and single data class makes it impossible for large-scale furnishing. We have brought the Internet of things (IoT) into the field of environmental obstacle to eject the problems in ritual processes. This paper is intended to implement a car emission surveillance scheme using the Internet of things (IoT), a green thumb for tracking down vehicles that cause taint on town highways and measure multiple types of toxic waste and their air level. This article presents at any moment anywhere using gas sensor a kind of real-time air pollution monitoring scheme. The measured information is communicated by text message to the car owner and domestic environment organizations. This assay demonstrates that the system is consistent, cost effective and can be tractably controlled, it can smell the car exhaust in real-time, and it can enhance the exhaust surveillance system's detection level and precision. This scheme offers excellent results only in urban regions in tracking air pollution. The main objective of smart emission monitoring system is to make it more innovative, user friendly, time saving and also more efficient than the existing system. Using smart systems not only efficiently takes an advance in environmental quality, but it also helps vehicle owner to save a lot of unnecessary troubles compared to the traditional emission test. G. Sarella et al. [16] presents the automated air pollution detection system for vehicles. This is intended to use semiconductor detectors in vehicle emission outlets that detect pollutant levels and also indicate this amount by a meter. When the amount of pollution/emission shoots beyond the threshold level already set, there will be a buzz in the car indicating that the limit has been violated and the car will stop after a certain period of time, a cushion moment provided to the driver to park his/her car. The GPS begins to locate the closest service stations during this time span. The fuel provided to the engine will be cut off after the timer runs out and the car will have to be transported to the mechanic or the closest service station. A microcontroller monitors and controls the synchronization and execution of the whole process. This idea, when augmented as a real-time project, will benefit the society and help in reducing the air pollution. The semiconductor sensors were used to identify the vehicle's contaminant amount. This concept is primarily focused on three blocks; smoke detector, microcontroller and injector of petrol. The smoke detector continually detects pollutants (CO, NOx, etc.). The microcontroller compares the level of pollutants with the stipulated level allowed by the government. When the pollutant level exceeds the standardized limit, it sends a signal to the fuel injector. On receiving a signal from the

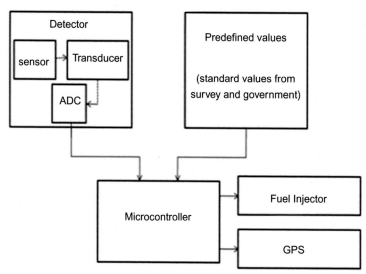

Fig. 3 Detector module.

controller, the fuel injector stops the fuel supply to the engine after a particular period of time. The overall block diagram of the proposed system is given in Fig. 3.

L. Myllyvirta and S. Dahiya [2], National Air Quality Index (NAQI), state that air pollution has become a major problem and a big threat to the public health and hazardous to the environment. It has been stated that more than 3 million deaths has been recorded over the year 2012 by World Health Organization (WHO) and more than 6 lakhs of deaths in India among them by estimated by global burden of diseases (GBD). After all this, NAQI was introduced in India. NAQI measures the pollution level in different areas and informs us about it. It generated its own data by using IoT techniques. Realizing the need and importance of public health, Central Pollution Control Board has started NAQI which plays major role in creating awareness in people (Fig. 4).

NAQI concludes that however there is lot amount of data available for 16 cities all over India and more than 50 cities are suffering by air pollution because of industrial clustering effecting more than a million people and hence major steps has to be taken to prevent pollution (Fig. 5).

Dr. S. Guttikunda [5] explains about the outdoor air pollution spreading very fast across many cities in both developing and developed countries. In both type of countries, cities only differ in the air quality index values and obviously AQI is high in developing countries because of large urban industries.

Number	Remark	Health Impact
1–50	Good	Minimal impact
51–100	Satisfactory	Minor breathing, discomfort to sensitive people
101–200	Moderate	Breathing discomfort to the people with lungs, asthma and heart diseases
201–300	Poor	Breathing discomfort to most people on prolonged exposure
301–400	Very Poor	Respiratory illness on prolonged exposure
401–500	Severe	Effects healthy people and serious impacts to those with existing diseases

Fig. 4 AQI pollution level and health impact by NAQI.

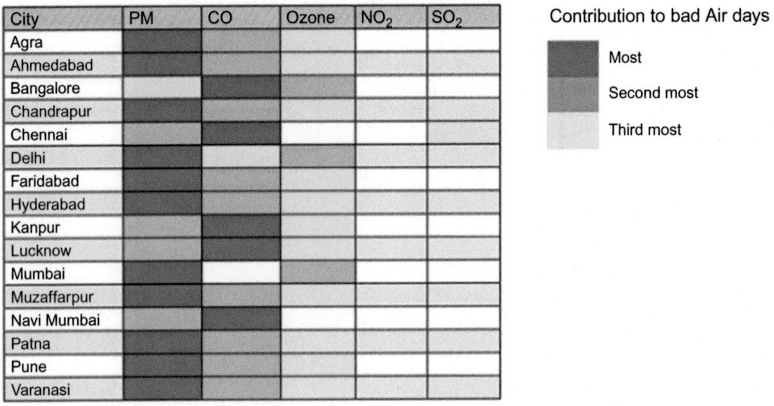

Fig. 5 Different pollutants contribution to bad air quality in different cities by NAQI.

This is because developing countries are concentrating more on rapid industrialization for growth of their country. The air pollutants arise from different sources especially from combustion sources, industrial outlets and automobiles. Nitrogen oxides (NOx), carbon monoxide (CO), carbon dioxide (CO_2), volatile organic compounds (VOCs) and particulate matter (PM) are the most important pollutants released into atmosphere by traffic. Other pollutants containing sulfur arise mainly form the industries. Among all these, PM accounts for more health hazards as it can cause chronic and many acute respiratory diseases. Automobile exhaust, industrial outlet of flue gases are main causes of release of PM into environment. The author explains that it's not only important to control pollution, but also to control the emission of pollutants into environment.

Finally, author concludes that along with public health, even environmental regulations are more effected in a very bad manner. As a part of pollution control, following regulations and creating awareness plays major roles.

Fuel	Advantages	Disadvantages
Electricity	Potential for zero vehicle emission.	Current technology is limited
Ethanol	Very low emission of ozone-forming hydrocarbon and toxics	High fuel cost
Methanol	Low emission of ozone forming hydrocarbon and toxic substance	High fuel cost
Natural gas	Can be made from variety of feed stokes Very low emission of ozone forming hydrocarbons, toxics ,and carbon monoxide.	High vehicle cost
Propane	Somewhat lower emission of ozone forming hydrocarbon and toxics.	No energy security or trade balance benefit.

Fig. 6 Use of different fuels.

S. Kumar and D. Katoria [3] explain about the air pollution and its control measures, author defines air pollution as presence of any foreign material in the air in excess quantity. Pollution has become a global challenge effecting human health and environment and causing serious threats for social well-being. Government and industries has to follow some strict measures toward pollution control. Author says complains that fuel combustion is the most important phenomena causing air pollution. So the author suggests how to control the exhausts of the fuel combustion, mainly CO and particulate matter (Fig. 6).

The author concludes that there are numerous ways of controlling air pollution and using different fuels in automobiles is one type. Author also represented various technologies embedded in industries like filtration of flue gases before they are released to atmosphere using filter bags and scrubbers.

3. Design

Architectural design is the process of defining a collection of hardware and software components and their interfaces to establish the framework for the development of a system. It defines the structure, behavior, and more views of a system (Fig. 7).

The proposed system is lightweight and compact on automobiles. This surely helps in reduction of harmful gaseous levels from vehicles. It provides a guarantee that it would be a very big change in the prevention and control measures for air pollution. The sensors and the other components used in building this system cost very less and hence the system is cost effective.

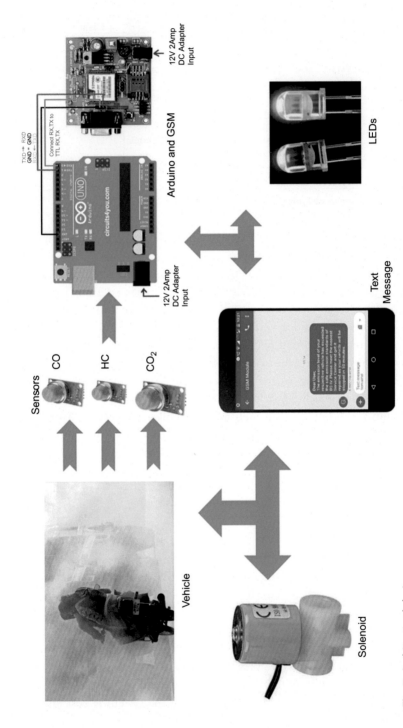

Fig. 7 Architectural design.

Apart from being cost effective, this system is high in performance as it uses low power consumption sensors and also for the fact that these sensors are highly precise in detection. If the value sensed by the sensors are higher than the threshold value, the fuel supply is instantly cut off preventing the vehicle to be mobile. A system architecture or systems architecture is the conceptual model that defines the structure, behavior, and more views of a system. An architecture description is a formal description and representation of a system, organized in a way that supports reasoning about the structures and behaviors of the system.

System architecture can comprise system components, the externally visible properties of those components, the relationships (e.g., the behavior) between them. It can provide a plan from which products can be procured, and systems developed, that will work together to implement the overall system. There are two different models in this project. The modules are as follows:

Module 1

Sensors are used to detect the pollutants released by the vehicle (Fig. 8). We use three different types of sensors, namely, MQ7, MQ2, MQ135 to collect different types of emissions. MQ7 is highly sensitive to carbon monoxide. MQ2 is suitable for detecting gas leaks. MQ135 can be used to detect NH_3, NO_x, alcohol, benzene, smoke, CO_2, etc. The value collected by each sensor is gathered at the Arduino board. Arduino has been programmed in such a way that it collects data from different sensors and compares it with standard values. There are six LED's used to depict the

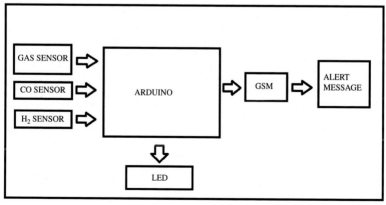

Fig. 8 IoT module.

current status of different pollutants from which three of them are red and are green. If the values received by the sensors are greater than the standard values the LED lights go red. If the values are with in standard values then the LED lights remain green. If the pollutant values remain to be greater than their standard value, Arduino passes the data onto the GSM module. GSM module is responsible for sending alert messages in order to notify the user.

The figure above depicts the architecture model of the hardware. The sensors are connected to the Arduino board. Arduino board constantly receives values from the sensors. All the LED lights are connected to the Arduino board through a circuit. The GSM module and Arduino board are connected via an interface. A mobile sim card is placed inside GSM module. The sim card is used to send different alert messages to the owner of the vehicles.

Model 2

The values gathered at the Arduino board are stored in a dataset. This helps us keep track of the changes happening in the air at different points of time. Once there are enough values present in the dataset, the data gathered can be used to create a model that can make analysis and predictions (Fig. 9). The figure above represents a model designed using machine learning algorithms. The dataset consists of real-time data which has been collected from different sensors. The dataset acts as an input to our model.

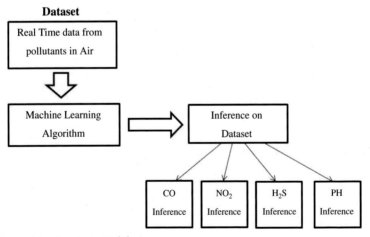

Fig. 9 Machine learning module.

The model uses a machine learning algorithm (decision forest regression algorithm in this case) to learn the different values that can be present in the dataset.

Once learning is done the model is now prepared to draw inferences on the different pollutants present in the air as well as do predictions of pollution level in future.

3.1 Graphical user interface

The main graphical user interface is made using a package available in R Studio called TclTk. Tcl stands for tool command language. It is a very powerful but easy to learn dynamic programming language, suitable for a very wide range of uses, including web and desktop applications, networking, administration, testing and many more. Open source and business-friendly, Tcl is a mature yet evolving language that is truly cross platform, easily deployed and highly extensible. Tk is a graphical user interface toolkit that takes developing desktop applications to a higher level than conventional approaches. Tk is the standard GUI not only for Tcl, but for many other dynamic languages, and can produce rich, native applications that run unchanged across Windows, Mac OS X, Linux and more. In Fig. 10 we can see a start button and a stop button which are used. When start button is clicked the sensors start detecting pollutants. Once we click on stop button the emission rate is displayed on the screen. Locate me button can be used to realizer the pollution level in the area around you. Close button quits the application.

4. Implementation and results

The work overall has been done in three different modules. The three modules are IoT module, data analytics module and Android app module. The overall functioning of the project is as follows:

The sensors, i.e., MQ-2, MQ-7 and MQ-135 sense the emission levels from the vehicular exhaust. These values are then sent to the Arduino for processing. If the emission levels are under control or hasn't crossed threshold, i.e., all three green LEDs will glow, the system will again collect the next set of values. If the threshold has been crossed, i.e., all red LEDs with glow, then a message is sent to the owner of the vehicle about the extra

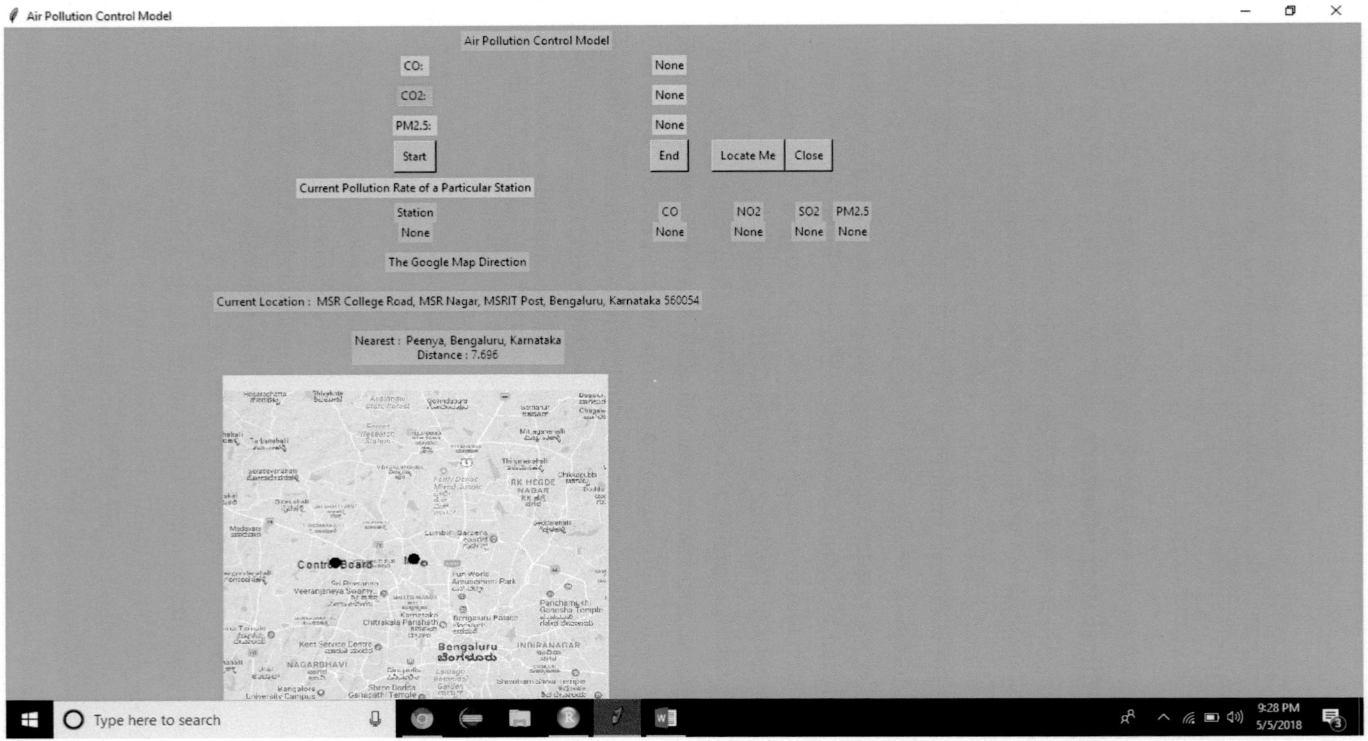

Fig. 10 GUI of the model.

emission level saying that if issue not resolved immediately the vehicle will be stopped in 10 min. Even after this, if his vehicle is still emitting extra emissions then his fuel supply is blocked by the use of solenoid valve.

4.1 Module description

Sensors: MQ7, MQ2, MQ135 are the different sensors used. MQ7 is a simple to use carbon monoxide sensor. It can detect CO concentration from 0 to 2000 ppm. MQ2 is a used for detecting gas leakages. It is very helpful for detecting H_2, LPG, CH_4, CO. MQ135 is an air quality sensor which detects wide range of gases.

Arduino: Arduino acts as a microprocessor, i.e., all data preprocessing occurs at the Arduino. It takes data from the sensors and compares the values with standard values. It updates the status by changing the condition of the LED lights. If the values are high it sends data to GSM module.

GSM: GSM is a digital mobile telephony system. It is used to send alert messages to the owner of the vehicle. It acts as an interface between Arduino and the user. Arduino sends the pollutant information to the GSM module. The GSM module uses this data and notifies the user.

Machine learning model: Decision forest regression is a technique used for training the model. A large portion of the air pollution data gathered is given as input for the model to learn. After completion of learning phase the model is tested with the remaining data. The test depicts the accuracy of the model. The model is now ready to perform predictive analysis for any new data (Figs. 11 and 12).

4.2 Actual data vs predicted data

Accuracy = mean of predicted data − mean of actual data
  Accuracy = abs (0.07344472 − 0.07094936)
  Predicted accuracy = 0.002495359

$$MAE = \frac{\sum_{i=1}^{n} |y_i - x_i|}{n}$$

where
  MAE = Mean absolute error
  Y_i = Predicted value
  X_i = Actual value
  n = no. of samples
  MAE = 0.002495359

Row	Actual_CO
82000	0.063786
82002	0.059671
82003	0.05144
82004	0.063786
82006	0.065844
82011	0.053498
82013	0.053498
82015	0.032922
82018	0.076132
82027	0.061728
82030	0.142546
82033	0.066042
82037	0.057613
82038	0.037037
82040	0.088477
82042	0.069959
82046	0.084362
82047	0.082305
82048	0.022634
82049	0.069959
82050	0.059671
82051	0.053498
82052	0.059671
82054	0.059671
82058	0.039095
82060	0.059671
82062	0.041152
82063	0.057613
82075	0.061728

Fig. 11 Sample of actual CO emission.

$$\text{RMSD} = \sqrt{\frac{\sum_{t=1}^{T} (\widehat{y}_t - y_t)^2}{n}}$$

where

RMED = Root mean squared error

$\hat{y}$ = Predicted value

y = Actual value

n = no. of samples

T = tuple

RMSD = 0.06188289

Row	Predicted_CO
82000	0.058448
82002	0.050152
82003	0.058404
82004	0.061571
82006	0.066013
82011	0.070858
82013	0.063261
82015	0.058845
82018	0.066795
82027	0.068744
82030	0.082279
82033	0.082279
82037	0.070897
82038	0.068074
82040	0.081954
82042	0.062502
82046	0.071635
82047	0.056945
82048	0.060428
82049	0.05592
82050	0.057738
82051	0.072202
82052	0.062976
82054	0.05526
82058	0.056665
82060	0.057874
82062	0.043161
82063	0.064508
82075	0.064079

Fig. 12 Sample of predicted CO emission.

4.3 Explanation of algorithm

Algorithm: To read the sensor data and process it.

Input: Sensor value, owner's phone number.

Output: Intimation in terms of SMS, Blocking of fuel supply.

Method:

```
Function air_pollution_control() {
Hydrocarbon gases(HC) = values_read_by_MQ2
Carbon Monoxide(CO) = values_read_by_MQ7
Carbon dioxide(CO2) = values_read_by_MQ135
```

```
Threshold_for_MQ2 = 400;
Threshold_for_MQ7 = 300;
Threshold_for_MQ135(in ppm) = 500;
Set Solenoid Pin to High;
Print values_read_by_MQ2, values_read_by_MQ7, values_read_by_MQ135
counter=0;
If (HC > Threshold_for_MQ2) {
        Red_LED_for_MQ2 set to High;
        Green_LED_for_MQ2 set to Low;
        counter=counter+1;
        }
If (CO > Threshold_for_MQ7) {
        Red_LED_for_MQ7 set to High;
        Green_LED_for_MQ7 set to Low;
        counter=counter+1;
        }
If (CO₂ > Threshold_for_MQ135) {
        Red_LED_for_MQ135 set to High;
        Green_LED_for_MQ135 set to Low;
        counter=counter+1;
        }
else {
        Red_LED_for_MQ2 set to Low;
        Green_LED_for_MQ2 set to High;
        Red_LED_for_MQ7 set to Low;
        Green_LED_for_MQ7 set to High;
        Red_LED_for_MQ135 set to Low;
        Green_LED_for_MQ135 set to High;
        }
If(counter>=3) {
        SendMessage();
        wait(10 minutes);
        Set Solenoid Pin to Low;
        }
}
```

One graph shows average values of the particular pollutant in all s stations, and all such graphs are combined (Fig. 13).

This graph shows the overall average values of the pollutants in every station all over Bangalore (Fig. 14).

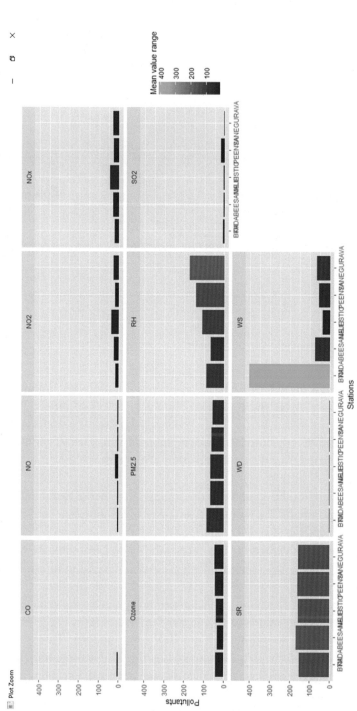

Fig. 13 Average values of pollutants in each station.

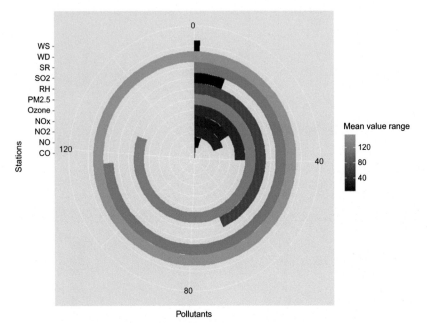

Fig. 14 Overall average values of pollutants.

This graph shows the overall average values of the pollutants in every station all over the Bangalore shown in points of different colors (Fig. 15).

Arduino IDE is the interface used for implementing the IoT project. The code given here is for the implementation of gas sensors, GSM module and the solenoid valve (Fig. 16).

The screenshot shown above is the configuration of the Arduino IDE for the emission system. The Arduino board used is Arduino Uno and the port used is COM3 (Fig. 17).

The screenshot depicts the interface of the Cool Term software which is used to collect the data obtained from the Arduino serial monitor into the text data (Fig. 18).

The screenshot shows the notification which is sent to the vehicle owner when the threshold value crosses the standard values (Fig. 19).

The screenshot show the Arduino IDE serial monitor. This is used to display the level of the pollutants, sensed by the MQ-sensors (Fig. 20).

The screenshot tells that the vehicle is going to be stopped in few minutes since the threshold value is crossed (Fig. 21).

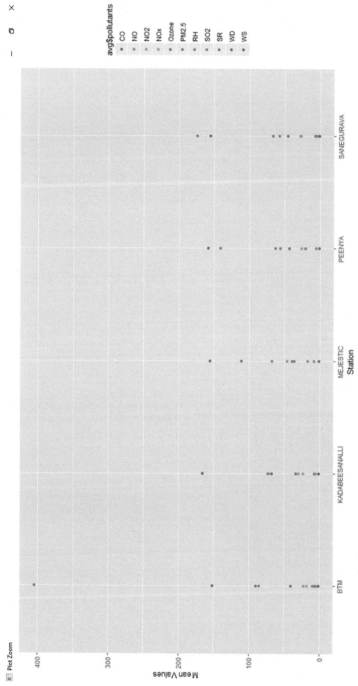

Fig. 15 Average values of pollutants in each station.

```
f1 | Arduino 1.8.5
File Edit Sketch Tools Help

f1

#include <SoftwareSerial.h>

SoftwareSerial mySerial(9, 10);


int redLed_2 = 2;
int greenLed_2 = 3;
int redLed_135 = 5;
int greenLed_135 = 4;
int redLed_7 = 7;
int greenLed_7 = 6;
int k1=0;
int smokeA0_135 = A4;
int smokeA0_2 = A5;
int smokeA0_7 = A3;
int count=0;
int solenoidPin = 12;      //This is the output pin on the Arduino we are using

int sensorThres_2 = 300;
int sensorThres_135 =300;
int sensorThres_7 =300;
void setup() {
  pinMode(redLed_2, OUTPUT);
  pinMode(greenLed_2, OUTPUT);
  pinMode(smokeA0_2, INPUT);

  pinMode(redLed_135, OUTPUT);
  pinMode(greenLed_135, OUTPUT);
  pinMode(smokeA0_135, INPUT);

  pinMode(redLed_7, OUTPUT);
  pinMode(greenLed_7, OUTPUT);
   pinMode(smokeA0_7, INPUT);
    pinMode(solenoidPin, OUTPUT);
     digitalWrite(solenoidPin, HIGH);
      mySerial.begin(9600);
   Serial.begin(9600);
}

void loop() {
   int analogSensor_2 = analogRead(smokeA0_2);
 // delay(400);
    int analogSensor_135 = analogRead(smokeA0_135);
    //delay(400);
```

Fig. 16 Arduino IDE interface.

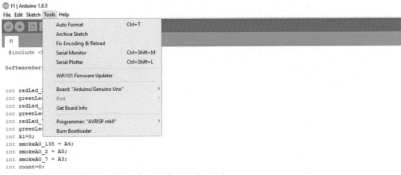

Fig. 17 Arduino IDE settings for Arduino board.

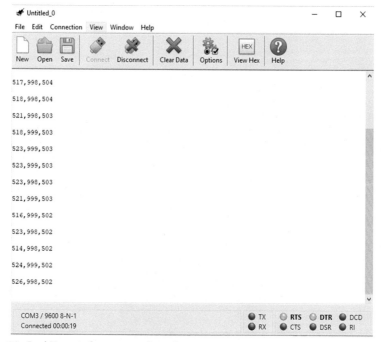

Fig. 18 Cool Term software to collect data.

The above graph depicts the error rate generated for individual trees in a random forest. As the number of trees increase we can see reduction in the error rates (Fig. 22).

This diagram depicts the prediction error realized during the testing process after tuning the parameters of random forest algorithm (Figs. 23–25).

5. Conclusion

Over the last few decades there has been an increase in the rate of pollution, leading to several environmental issues. There will be an enormous population that does not take the pollution from their cars seriously, which has already caused several environmental issues like depletion of the ozone layer and so on. So, this scheme is going to be very useful to curb this issue. Smart emission surveillance system's primary goal is to create it more innovative, user friendly, time saving and also more effective than the current system. Using smart systems not only takes a step forward in environmental quality, it also enables car owners save a lot of unnecessary problems

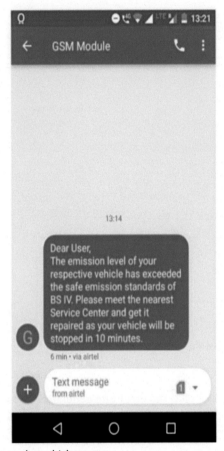

Fig. 19 Notification to the vehicle owner.

compared to traditional emission testing. The idea of identifying and indicating pollution level to the driver. This scheme only monitors three parameters and can therefore be extended by considering more parameters that cause the cars in particular to pollute. This system gives availability of viewing the sensor outputs through internet. It can be done by providing orders from a distance to regulate emissions. Many pollutants do not have detectors which, if available, are very costly and therefore constructing detectors for distinct parameters could be a very difficult job in the future. The fact that this system is just an add-on, as it does not change the configuration of the engine by any means, will make it easier to employ this system in the existing vehicles. The same concept can also be extended to industries.

Fig. 20 Arduino IDE serial monitor.

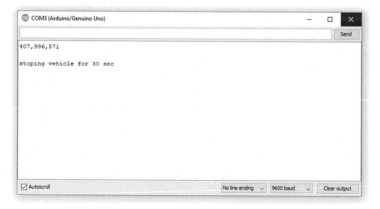

Fig. 21 Arduino IDE serial monitor with stopping message in terminal.

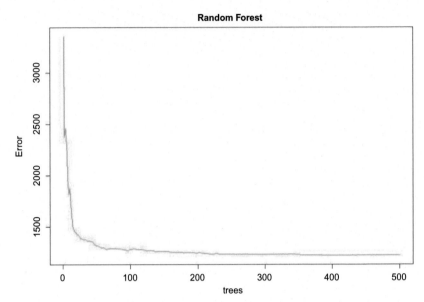

Fig. 22 Model error rate decreasing with increase in number of trees.

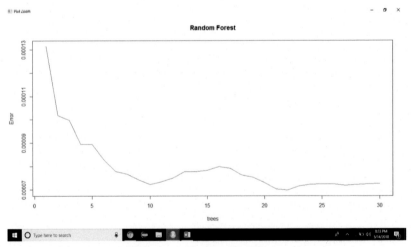

Fig. 23 Model error rate decreasing with trees (0 − 30).

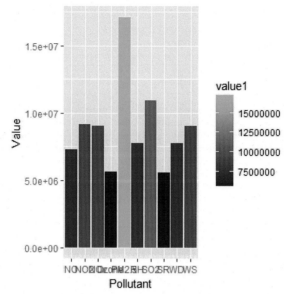

Fig. 24 Feature importance histogram.

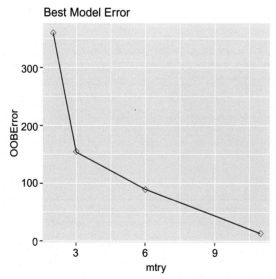

Fig. 25 OOB error rate.

References

[1] Ghewari MU, Mahamuni T, Kadam P, Pawar A: Vehicular pollution monitoring using IoT, *Int Res J Eng Technol* 05(02).

[2] Myllyvirta L, Dahiya S: *A Status Assessment of National Air Quality Index (NAQI) and Pollution Level Assessment for Indian Cities, 2015:* 2015, Greenpeace India greenpeace. org/india. https://www.greenpeace.org/india/Global/india/2015/docs/India-NAQI-PRESS.pdf.

[3] Kumar S, Katoria D: Air pollution and its control measures, *Int J Environ Eng Manag* 4(5):445–450, 2013. https://pdfs.semanticscholar.org/ed3f/1297ea9b3576c592587a6 960ce198daa9d0a.pdf.

[4] Rose Sweetlin T, Priyadharshini D, Preethi S, Sulaiman S: Control of vehicle pollution through Internet of things (IOT), *Int J Adv Res Ideas Innov Technol* 3(2) Available online at, www.ijariit.com.

[5] Guttikunda S: Air Quality Index (AQI): Methodology & Applications for Public Awareness in Cities, 2010: http://www.urbanemissions.info/wp-content/uploads/docs/SIM-34-2010.pdf.

[6] Kspcb: http://kspcb.kar.nic.in/.

[7] Hunshal B, Patil D, Surannavar K, Tatwanagi MB, Nadaf SP: Vehicular pollution monitoring system and detection of vehicles causing global warming, *Int J Eng Sci Comput* 7(6):12611, 2017.

[8] https://en.wikipedia.org/wiki/Arduino.

[9] https://en.wikipedia.org/wiki/PyCharm.

[10] Lolge SN, Wagh SB: A review on vehicular pollution monitoring using IoT, *Int J Electr Electron Eng* 9(01):745, 2017.

[11] Usha S, Naziya Sultana A, Priyanka M, Sumathi: Vehicular pollution monitoring using IoT. In *National Conference on Frontiers in Communication and Signal Processing Systems (NCFCSPS '17),* vol. 6, 2017, An ISO 3297: 2007 Certified Organization. Special Issue 3.

[12] Bangal SP, Gite Pravin E, Ambhure Shankar G, Gaikwad Vaibhav M: IoT based vehicle emissions monitoring and inspection system, *Int J Innov Res Electr Electron Instrum Control Eng* 5(4):410, 2017, ISO 3297:2007 Certified.

[13] https://en.wikipedia.org/wiki/GSM.

[14] https://www.sparkfun.com/products/9403.

[15] Smruthie S, Suganya G, Gowri S, Sivaneshkumar A: Vehicular pollution monitoring using IoT, *Int J Digit Commun Netw* 2: 2014.

[16] Sarella G, Khambete AK: AMbient air quality analysis using air quality index—a case study of Vapi, *Int J Innov Res Sci Technol* 1(10):21319, 2015.

Further reading

[17] http://wiki.seeedstudio.com/Grove-Gas_Sensor-MQ2/.

[18] https://www.olimex.com/Products/Components/Sensors/SNS-MQ135/resources/SNS-MQ135.pdf.

[19] https://www.rhydolabz.com/sensors-gas-sensors-c-137_140/air-quality-sensor-mq135-p-1115.html.

[20] http://events.awma.org/files_original/ControlDevicesFactSheet07.pdf.

About the authors

Chetan Shetty is working as a Lead Data Scientist, HCL Technologies, Bangalore. His area of interest includes Machine Learning and Deep Learning.

B.J. Sowmya is working as an Assistant Professor, Department of Computer Science and Engineering, M S Ramaiah Institute of Technology, Bangalore 560054. Her area of interest is Data Analytics, Machine Learning, and Internet of Things.

Dr. S. Seema is working as Professor, Department of Computer Science and Engineering, M S Ramaiah Institute of Technology, Bangalore 560054. Her area of interest is Data Analytics, Machine Learning, and Internet of Things.

Dr. K.G. Srinivasa is working as Professor at National Institute of Technical Teacher Training & Research, Chandigarh, Chandigarh, India. He is the recipient of All India Council for Technical Education— Career Award for Young Teachers, Indian Society of Technical Education—ISGITS National Award for Best Research Work Done by Young Teachers, Institution of Engineers(India)—IEI Young Engineer

Award in Computer Engineering, Rajarambapu Patil National Award for Promising Engineering Teacher Award from ISTE—2012, IMS Singapore—Visiting Scientist Fellowship Award. He has published more than hundred research papers in International Conferences and Journals. He has visited many Universities abroad as a visiting researcher. He has visited University of Oklahoma, USA, Iowa State University, USA, Hong Kong University, Korean University, National University of Singapore are few prominent visits. He has authored two books namely File Structures using C++ by TMH and Soft Computer for Data Mining Applications LNAI Series—Springer. He has been awarded BOYSCAST Fellowship by DST, for conducting collaborative Research with Clouds Laboratory in University of Melbourne in the area of Cloud Computing. He is the principal Investigator for many funded projects from UGC, DRDO, and DST. His research areas include Data Mining, Machine Learning and Cloud Computing.

CHAPTER NINE

The human body: A digital twin of the cyber physical systems

Janet Barnabas[a], **Pethuru Raj**[b]
[a]Department of Computer Applications, National Institute of Technology, Tiruchirappalli, India
[b]Reliance Jio Infocomm. Ltd. (RJIL), Bangalore, India

Contents

Advances in Computers, Volume 117
ISSN 0065-2458
https://doi.org/10.1016/bs.adcom.2019.09.004

Abstract

The human body is an excellent engineered product from which many computing systems have drawn its inspiration. The seamless integration of computation and physical components create and substantiate the human body as an excellent parallel of the cyber physical systems (CPS).

The human body is a combination of computation, communication and control processes. The Embedded sensors all over the body monitors and controls the physical processes, with feedback loops where physical processes affect actions and reactions and vice versa. This system can be termed as a digital twin of the present CPS.

In this chapter, we will analyze the architecture of a cyber physical system and spring parallels to the functioning of various systems in the human body to gather pearls of wisdom from its design as it is a multifarious digital twin of the CPS. The twin is biological systems of the human body and not a virtual system.

1. Introduction

The terminology used for most computer technologies are borrowed or inspired from nature and creation. The systems in the world can be classified broadly as physical and virtual systems (Fig. 1). The living systems include social, organism and ecosystems. The social systems include family, school, village, community to name a few. The organism includes animals, plants, microorganisms, human beings. The ecosystem includes forest, pond, lake, sea etc. The nonliving or manmade systems are mechanical, optical, electrical and cyber systems. All these systems work based on some laws and have interactions with each other. The manmade systems derive their names from the type of energy they use. The cyber systems are cyber physical, virtual or social systems.

The cyber physical systems (CPS) is an inspiration from the human systems. They are smart devices with embedded sensors, actuators and processors that are networked with each other to interact with the physical world, and supply real-time data of the performance of these devices in specific

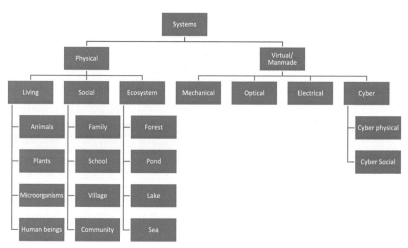

Fig. 1 Hierarchy of systems.

applications. CPS have transformed the structure of working of the industry and cyber world, by incorporating smartness and proactive usage of the available resources on a wide range of applications thus pervading into the lifestyle of the modern human being. In short, CPS has made the manufacturing industry more thriving with transformative visions expanded from constant monitoring of sensor and actuator data that is made accessible and comprehendible.

The sensors that are linked to real-time systems stream in data continuously. This valuable data are converted to information and it is crunched to produce knowledge. This knowledge pattern is deciphered to make meaningful decisions. The decisions are made to control the sensors and make them work smarter based on feedback.

The human body is an excellent twin of CPS. There are 11 distinct human body systems. They are digestive, cardiovascular, immune, endocrine, muscular, integumentary, reproductive, nervous, skeletal, respiratory, and urinary systems.

These systems work in perfect synchronization and incrementally learn as they grow to resolve problems. They accumulate a lot of data and use it to effectively function in day to day life. The data are stored centrally in the brain and also in the nerves to help with reflux action. The system works in perfect synchronization and converts the information into wisdom for better decision making. This data is also shared and acquired from others to reinforce the working of the system. Rarely, external influence is needed to streamline the system by using medicines or treatment. But mostly, given time and rest, the systems recover faster. When they are abused, the system does not function properly. Then it needs external help to come back to its normal state.

The various systems of the body are like the cyber physical systems in the world. They are interconnected and dependent on each other. With the advent of Internet of Things, all the devices became smarter. As these smart devices which are interconnected work together, a lot of data is created. This created data is analyzed using deep learning algorithms. It has led to data driven decision making where large amount of data trains the system to help substantiate previously unthinkable decisions. Now the industry 4.0 tried to integrate this model into their working to make the shop floor smarter. The availability of internet bandwidth paved the way for connection of everything. So Internet of everything was integrated into the industry to make it cleverer. The number of accidents were drastically reduced. The manpower required was lesser and productivity increased.

Thus, the cyber physical system has revolutionized the world and made previously unthinkable and dangerous missions possible with the power of connectivity and data.

2. Cyber physical systems

In CPS, we have the computing, control, sensing and networking components that every system should have. It is an offshoot of Internet of Things (IoT), Internet of Everything, Industry 4.0 [1], Machine-to-Machine (M2M), TSensors (Trillion Sensors) [2,3] invented as autonomous cars, biometric screening, wearables, AI-enhanced workplaces and smart living spaces, smart cities, edge computing and many more. These also abuse the smart sensors and triggers the next stage of innovation toward intelligent and autonomous systems.

It is an intermixing of systems and processes that are complex in itself and still interact with each other. The interaction is facilitated by the computation, communication and control which are the three C's of the various technologies backed up by the intelligence generated by deep learning models.

The CPS has these three important components that define the entire system in an industry.

2.1 The physical layer

Sensors, actuators, processors and transceivers form the base in the lowest layer of the CPS architecture used for data gathering. The continuous and streamed data from this lowest layer are stored in a cloud or fog or edge. The data are valuable information for the next layers to process in the CPS architecture.

The computation part of the layer uses the data for analysis to garner useful information about the system and maintain the health of the CPS. The smart sensors do not transmit all data but only those that are needed for informed decision making.

The data collected from the sensors or actuators is very humongous and varied. This data has to be aggregated and preprocessed into digital streams to enable data processing. To carry out this data processing, it is imperative that a data acquisition system (DAS or DAQ) be used. Data acquisition is the process of measuring real world physical conditions using sampling signals and converting the resulting samples into digital numeric values that can be manipulated by a computer. Data acquisition systems typically convert analog waveforms into digital values for processing. The data that is collected is collated to be sent to the data processing center so that it can be used for learning, generating intelligence information and making informed decisions. It has to be visualized for quick understanding of the data [1]. Data visualization is another field of interest as it helps to understand the large data clearly and try to find patterns and trends in them. It helps to isolate anomalies easily and finds correlations between sets of the data.

2.2 The network layer

Earlier all the systems were independent and could not communicate with each other. Human beings formed the mode of communication between them. For example, consider a washing machine. Earlier we had semi-automatic machines that needed human intervention to put the clothes into the drier. Later, the machine was integrated with sensors and actuators to use a same drum for washing and drying. The scenario has changed to a network of systems where they seamlessly communicate with each other. Machine to machine communications happen effortlessly today. Machine instructs other machines to work and makes decisions.

The DAS connects to the sensor network, aggregates outputs, and performs the analog-to-digital conversion. The Internet gateway receives the aggregated and digitized data and routes it over wired or wireless

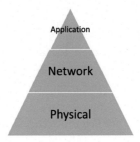

Fig. 2 Architecture of CPS.

connections, to next layer for further processing. So this layer basically deals with communication between smart devices. The communication is wired or wireless depending on the device and application. The important issues in this layer are security, privacy and trust.

2.3 The application layer

This layer controls the smart devices based on the data received from the previous layers. The sensors will get feedback based on the intelligence data generated and proper control signals will be generated to control the health of the sensors and actuators or make possible changes in the behavior of the sensor according to changes in the external environment. This layer delivers application specific services to the user. Also the sensors can be made to work smarter based on the feedback from data driven decision making using deep learning models.

The CPS architecture has the above mentioned layers as depicted in Fig. 2. The sensor layer is very dense and a lot of data is collected here. It is then sent to the network layer where the data is analyzed and patterns mined. This is used by the application layer for decision making. All the layers work independently and are connected commonly by data between them.

3. Biological systems

There are 11 systems of the human body. All the systems work in close association with each other. Biologically, they are interrelated and intertwined with each other. The systems work in perfect harmony with each other. When one system is weak, it affects other parts of the body. A trained doctor is able to correctly find out the problem with the system as he is making a data driven decision based on his previous knowledge of the human biological system. He compares with its present state and locates the problem.

All the organ systems have a unique function, and each organ system also depends, directly or indirectly, on all or some of the other systems. Thus when one system is affected, it indirectly affects the other systems and results in shut down of all the systems meaning death of a human.

There are some vital organs like heart, lungs, liver, kidney and brain. These form the important part of the various systems as their driving force. These living organs make these systems work, integrate, provide feedback and gather data.

4. The systems of the human body

Ross Toro [4] provides a peek into the various systems of the human body in his blog. These systems work in tandem to help in the smooth functioning of the body. Each system is unique and has its own rules, specialty and functions. Amazingly, they all work together in perfect symphony to keep the system productive. Repairing, restoring, healing, supporting, reacting in response to the immediate need, these systems are the most complex example of the present day cyber physical system (Fig. 3).

Fig. 3 Subsystems of the human body.

4.1 The cardiovascular system

It includes the blood, the heart and vascular network. The functions of this system are to move nutrients and waste products through the body and assist with maintaining body temperature and pH. The heart pumps the blood through the body and maintains the pressure. Blood delivers its payload to the parts of the body and brings back the waste for cleaning. Thus the same system is used for two different functions through two different blood vessels making it cost effective and easy to maintain and monitor. The interlinked network of small blood vessels that make this transport of blood to parts of the body and cleaning of the blood is the important work of the vascular network. It is a wired connection where all the body parts are interlinked and provided with power and needed nutrients to grow and perform their duties. The blood also cleans up the system periodically to remove the unwanted waste from the cells.

4.2 The circulatory system

It consists of three independent systems that work together: the heart (cardiovascular), lungs (pulmonary), and arteries, veins, coronary and portal vessels (systemic). It is responsible for the flow of blood, nutrients, oxygen and other gases, and hormones to and from cells. It delivers oxygen and nutrients to organs and cells and carries their waste products away.

The system has huge blood vessels called arteries carrying pure blood that is pumped from the heart. The blood flows through the capillaries in the body to exchange oxygen and nutrients to the muscles and cells. The vein carries back the waste materials and carbon dioxide from the cells to be cleaned. Then, fresh blood is pumped again making the cycle complete. The blood is constantly replenished and the bone marrow helps in forming the blood cells.

The capillaries are small and thin structures that connect the artery and the vein. The blood flows through all parts of the body forming a complex network. When viewed from the cyber physical system perspective, it looks like a smart system that links all the parts of the body providing the much-needed energy and strength for performing their functions. They also clean up the system in the same way. They are not independent. They are heavily dependent on digestive system for the nutrients, lungs for oxygen, kidney for cleaning and artery and veins for transport through the body and capillaries for enabling the exchange process. They are delicate structures engineered creatively to perform these complex functions with utmost precision repeatedly, day in and day out.

4.3 The digestive system

It comprises of the gastrointestinal tract and the additional organs of digestion that include tongue, pancreas, salivary glands, liver, and gallbladder. Digestion breakdowns the food into smaller components, until they can be absorbed and assimilated into the body. Mechanical and chemical processes provide nutrients via the mouth, esophagus, stomach and intestines. It also eliminates waste from the body through the anus. This system provides energy rightly called power for the body to function.

The food is digested right from the mouth by the digestive juices that flow into the mouth. Saliva is produced by more than 1000 glands that generate them. Then, the food moves down the esophagus to the stomach where digestive acids break them down further. In the small intestine, the nutrient is absorbed into the blood. The pancreas, gall bladder and liver help in major part of the digestion process. They regulate insulin and break down food. These organs are near the stomach and drain their fluids into the stomach to aid in digestion. So it is not one single system working but a group of systems working to achieve the target of acquiring the nutrients needed for the upkeep of the body.

4.4 The endocrine system

It is a collection of glands that produce hormones that regulate metabolism, growth and development, tissue function, sexual function, reproduction, sleep, and mood, among other things. Salivary glands and sweat glands are some examples. They also provide chemical communications within the body using hormones. The hormones regulate many processes in the body and cause various diseases when not in sufficient supply or over supply. For example, if the hormone insulin is low, it results in high blood sugar. If it is high, then it results in low blood sugar. Thyroid hormone also is known to cause obesity if found in excess. The hormones estrogen and progesterone help in pregnancy and lactation. They are found to be present only in certain durations and for specific functions.

The hypothalamus connects the endocrine system with the nervous system. The pituitary gland is instructed by it to start or stop making hormones.

This pituitary gland uses the information from the brain to tell the other glands to make hormones according to the need of the body. The adrenaline is experienced by all of us in times of excitement or stress. It affects metabolism and sexual function. The endocrine system gets instruction from the brain that connects all other systems together. Thus the body functions are regulated by this system as and when it is required.

4.5 The immune system

It is the body's defense against infectious organisms and other invaders. The immune system attacks organisms and substances that invade body systems and cause diseases by a set of steps called immune response. The parts of the immune system are the tonsils and thymus, which make antibodies, the lymph nodes and vessels (the lymphatic system), Bone marrow, the spleen, which filters the blood by removing old or damaged blood cells and platelets and helps the immune system by destroying bacteria and other foreign substances and skin. The system also comprises of a network of lymphatic vessels that carry a clear fluid called lymph. It defends the body against pathogens that may endanger the body. It acts as the security mechanism. When a foreign object like thorn enters into the skin, the immediate response is pain, inflammation and raised temperature to remove the object. These happen immediately to locate and treat the breakdown. The bone in the toe if broken swells up to protect the breakage and fever results if left untreated. If given proper rest, it heals itself slowly. Thus the human immune system can discover, remember and decode pathogens. It can also destroy pathogens. If not able to, then it can isolate the area and seek external help by causing immense pain taking us to the doctor immediately. So this system is proactive and intelligent in securing the human body.

The food that we take boost the immune system. Garlic and turmeric are very good immune system boosters. They are digested by the digestive system and create enhanced immunity. This is external input to the immune system.

The CPS has a lot of takeaways from this. The robotic parts of the system must be continuously oiled and serviced to maintain the health of the system. If there is reduction in the level of oil, it has to be replenished immediately to prevent wear and tear of the machine. Periodically, external and human verification is needed to check for unreported problems as the system has no concept of pain. There is only replacement of worn out parts and not regeneration in CPS. So the health of the system is monitored continuously to replace at the right time.

4.6 The integumentary system

It consists of the skin, hair, nails, and exocrine glands. The skin is only a few millimeters thick. It is the largest organ in the body. The average person's skin weighs 10 pounds and has a surface area of almost 20 square feet. The functions of the skin are protection against toxins, radiation and harmful pollutants, generation of vitamin D from sun rays, excretion through sweat, and secretion of oil and melanin pigment for protection, regulation of body temperature by sweat evaporation and sensation of touch and feeling for social connectivity.

On the skin, if we feel the ant crawling, immediately our hand moves to remove it before it bites. This action is because the skin and hairs projecting from hair follicles in the skin, can detect changes in the environment. The change is transmitted to the central nervous system (brain and spinal cord), which respond by activating the skeletal muscles of eyes to see the ant and the skeletal muscles of the body to remove the ant [5]. The eyes see if the ant is dangerous. If so it is killed. Else it is removed. This process involves many sense organs and the nervous system. The speed of transfer of information and the reaction is instantaneous due to learning from the previous incident that causes us to fear and remember the pain associated with it.

The skin also maintains the body temperature and removes impurities through sweat. It protects the internal organs as a single large protective covering with a lot of sensors attached to them that periodically communicate with the other systems.

4.7 The muscular system

It consists of skeletal, smooth and cardiac muscles. It enables movement of the body, maintains the posture and circulates blood inside the body. It has over 600 muscles. Each muscle is a collection of cells that is attached to a single bone and is responsible for its movement. The muscle has two attachment points with the bone, one of which is moveable and the other fixed. The movement is by contraction, which is stimulated by nerve impulses and triggering the movement of the muscle, whereas relaxation happens when the impulse is removed and the muscle relaxes back to its natural state. The nervous system acts as a central control of the entire muscular framework controlling their movement. Thus no system acts of their own but is linked and dependent on one another for their smooth working. But the work of each component is clearly defined within the system.

4.8 The nervous system

The most complex system is the nervous system which consists of the brain, spinal cord, sensory organs, and nerves throughout the body that connect the brain to the organs. It coordinates the actions and sensory information by transmitting signals to and from different parts of the body. The nervous system detects environmental changes that affect the body, then works with the endocrine system to respond to such events. It collects and processes information from the senses via nerves and the brain and communicates so that the muscles contract to create physical movement. Thus it forms the network and communication layer of the systems. It can be compared to the electrical wiring in the electric systems.

The basic functions are to collect sensory input from the surroundings, enable motor movement and learn to make associations from previously stored information in the brain, to enable reflux action. Seeing a person will trigger the brain to identify and locate the person. Then, the face is instructed to smile at the person and the muscles in the face create a smile. This is a complex emotion created when the nervous system, sensory system, muscular system and skin work together.

4.9 The reproductive system

The genital system is a system of sex organs which work together to produce offspring. It keeps the generations alive and prevents extinction. The male and female systems are different and have different functions. The female system is used to create and nurture the offspring. The male system is used to deposit sperm to create the offspring. The function of the male and female system have been clearly defined to enable proper working and generation of offspring.

4.10 The respiratory system

The air enters the body through the nose or mouth. It passes through the sinus where its moisture and humidity is regulated and sent to the trachea which filters the air. Then it is sent to lungs through the bronchi which is lined up with cilia that move up and down to collect dust and germs in mucous that is expelled when we cough or sneeze. The basic functions are absorbing oxygen and breathing out carbon dioxide. This provides the required energy for the system and muscles to work. The system has its own mechanism of security to protect the organs from wear and tear and prevent invaders. The alveoli helps in exchange of the gases with the help of tiny capillaries, into the blood stream. The breathing motion is involuntary where it happens by automatic contraction and expansion of the muscle at the bottom of the chest called diaphragm.

For each part of the respiratory system, there are protection mechanisms that prevent the entry of foreign particles and infection. In CPS also we must look at the health of each part of the CPS machine and calculate the health factor of them.

4.11 The skeletal system

The adult skeleton consists of 206 bones, as well as a network of tendons, ligaments and cartilage that connects them. It performs vital functions like support, movement, protection, blood cell production, calcium storage and endocrine regulation that are needed for survival. Bones support the body and protect its internal vital organs. The skull protects the brain. The ribs protect the lungs. The external functions of the bone are for strength and to help in carrying heavy objects. The posture is a vital mechanism which when abused results in chronic pain. This posture is maintained by the skeleton system.

4.12 The urinary system

The renal system consists of the kidneys, ureters, bladder, and the urethra. The purpose is to eliminate waste from the body, regulate blood volume and blood pressure, control levels of electrolytes and metabolites, and regulate blood pH.

They remove the toxins in the body and keep the system clean. In this system, there is automatic maintenance every moment to remove the unwanted material from the body. This helps in maintaining the organs and keeps the system healthy.

All the above described systems are well researched and found to be working every day. The parallels drawn from these systems will result in a Bio-Plausible cyber system that can be created based on the insights gained from this biological creation.

5. Bio-plausible cyber systems

The layers of the cyber physical systems can be compared to the human systems described earlier. The physical layer where the computation happens is in the brain, endocrine system, immune system and integumentary

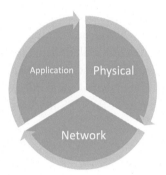

Fig. 4 Architecture of human systems.

systems. The network layer provides communication using the cardiovascular system, circulatory system, nervous system and respiratory system. The application layer provides control in the nervous system, reproductive system, muscular system, skeleton system and urinary system. The architecture of the human system is as shown in Fig. 4. All the systems work together and are interrelated in their applications.

The human body involves many biological systems. It provides automatic command and control of the smart systems inbuilt from scratch based on some coded genes called DNA. The systems of the human body when compared with the layers of the CPS are found to be highly complex and in the subatomic level. Though the technology has not reached the level of the biological sensors and actuators, we still will be able to mine valuable insights from human biology.

The Bio-plausible cyber system is differentiated from the cyber system by Ref. [6] where they try to establish a bio cyber physical system.

The Table 1 compares the biological systems with the cyber physical systems.

The Physical sensors of the human body are the sense organs like eyes, nose, ears, tongue, glands, nerves and skin. The actuators are muscles, joints, ligaments, synapses, valves in the heart, throat, eyelids, stomach etc. The processors are the brain and spinal cord.

The external stimulus plays a vital role in regulating the human system. The stimulus is touch of a hot object by a finger. The sensor is the temperature/pain receptor on the finger that senses it and relays it to the nervous system (spinal cord and brain), which is the coordinator. The coordinator makes the decision of how to react, and then commands the hand muscles (acting as the effector) to jerk back quickly. The framework takes us from stimulus (touch) to response (move hand away movement).

Table 1 Comparison of biological and cyber physical systems.

	Biological	Cyber physical
Physical layer	Sensors and actuators	Sensors, actuators, processors
Interactions	Nonlocal	Local and nonlocal
Symmetricity	No	Symmetric
Locality	Local	Local and global
Number of sensors	Many	Few
Speed of communication	Faster	Slower
Response rate	Fast	Slow
Communication	Across systems	Within systems
Storage	Online storage	Offline and cloud storage
Learning	Automatic and incremental continuous learning and Intelligent improvement	Data driven slow learning
Data processing	Streaming continuous data processing	Discrete data processing
Integration	Seamless integration of components	Connected components
Control	Centralized by the brain	Centralized or decentralized by cloud
Visualization	On sensors by reactions	On cloud

A similar stimulus is provided by the temperature sensor fitted in a smart room. As the temperature increases, it sends an alert to the edge device to analyze the temperature data. As it increases above a threshold, the cloud sends an instruction to the edge device that the air-conditioner in the room is to be turned on. The temperature is constantly monitored periodically to adjust the air-conditioner temperature to conserve energy (Fig. 5).

Though the two systems appear very similar there are many differences in their working patterns and methodology. The human system works in the micro level whereas the CPS works in the macro level. The level of interaction between the human systems is seamless and secure. It is learnable and trainable. It can be built incrementally whereas CPS is built once as a full grown system with specific macro objects in them. The individual systems

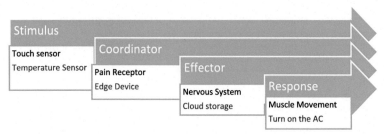

Fig. 5 The human request response system compared to CPS.

are made to interact with each other by human intervention or data driven automatic deep learned decision making. The mode of communication is wireless as wired communication restricts movement. The wireless mode results in increase in the vulnerability of the system as there is more chance of interruption and modification of the message resulting in loss of integrity and confidentiality of information.

A digital twin is a computational-relation-model of the physical entity which means that it can virtually replicate the behavior of the physical machine, and give an insight on how the machine will react when prompted with various actions.

The CPS can be digitally twined to provide framework for cyber physical social systems [7] which include smart cities, factories, schools, education institutions, agriculture, energy conservation and many more. Whatever is done manually can be automated with the help of CPS and artificial intelligence.

The cyber physical social systems include the human component in the architecture along with the physical, network and application layer. The social layer is responsible for intelligence gathering based on the human interaction with the system.

For example, if we consider the auto-driving car, the physical layer is the sensors in the car that collect real-time data. The network layer is the server or cloud where the data is crunched to form patterns or derive meanings. The application layer is where the decisions are made for smooth running of the car. But the social layer contains the nature of the passenger, capabilities, malicious intents, preferences, likes and dislikes of the people in the car. Some may want a short trip. Others may want to see the town. These kind of preferences are based on individuals and is called social characteristics. They are hard to encode and is variable in nature. Fig. 6 shows the cyber, social and physical layer of the CPSS.

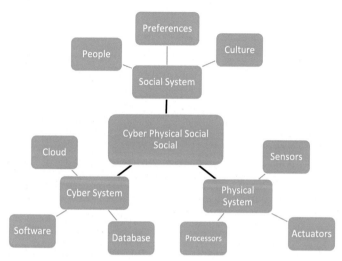

Fig. 6 Cyber physical social systems.

5.1 The domains of CPS

The various domains of concern of CPS when compared to human systems are Trust, Identity, Privacy, Protection, Safety and Security.

5.1.1 Trust

Each of the sensors must be able to trust the other sensors connecting remotely with them. In the human anatomy, every sensor of the body can easily identify a foreign object by their DNA footprint. Thus a high level of trust is built into the sensors. The identity of each sensor is established. Research in this domain is nascent as smart edge devices still have trust issues in identifying each other in a network. Also an untrusted component when entered into the human body causes pain. This type of reflux is not available in the cyber systems as the systems are not living and hence do not cause pain. A Parameter invariant monitor design (PAIN) Monitor has been discussed in research papers [8] given the need for safety by pain detection and recovery. The problem faced real-time in health care is the constant false alarm generated as the system is very sensitive to changes. It results in the system being shut down or ignoring the problem itself.

5.1.2 Identity

The cells in the body are identified by identity tags called Antigens. These identify the cells as a part of tissue or organ. It helps in identifying and preventing illegal entry of invaders and keeping the human system secure.

Thus CPS also must define an identity tag to secure the system where each edge device or sensor will be able to uniquely identify each sensor connected to it. Now we assign unique sensors to the devices thereby identifying each sensor. Sensors in a network must be given similar identity tags to identify themselves as a part of the peer network.

5.1.3 Privacy

Privacy is a domain where data is kept private for individual usage where no third party can access it without conscious revelation. The human mind is a perfect example of protection of the privacy of information. The brain stores experiences and retrieves them based on a query given to it. If a face is seen, then the name of the person and the park where we first met them is retrieved and presented for us to correlate. Thus the data are stored as a collection of interrelated experiences by the brain facilitating easy retrieval [9,10]. When data are interrelated, clustering techniques help to store data better and retrieval as a cluster increases semantic meaning.

5.1.4 Protection

There are physical, chemical, hardware and software protection mechanisms in the human body. The physical protection is the skin visible outside to all the world. Also the skull protects the brain, nails protect the fingertips. The white blood cells protect the body from pathogens. The hair in the nose protects from dust particles. The nails protects the fingers. Thus each of the organ is clustered together to provide a cluster of sensors that work together for a common purpose. For example, the sensitive skin of the fingertip is used to feel and also do hard work. It is possible as the sensors in the skin are active and clustered in the tip for proper working. The human body regulates the blood flow and monitors the working of the body to make it productive. The changes in external parameters affect the body. So it instructs the feet to walk to a tree when its sunny. These instructions protect the system and make it efficient.

In the CPS, the sensors and actuators must be clustered into groups and a group of sensors must be analyzed to prevent false positives. Also clustering must be based on nature inspired algorithms like ant colony or social spiders optimization [11,12] to make it as similar as possible to natural clustering mechanism. The human body works on these clusters of sensors that are used to reinforce a decision and give feedback to the system continuously. Problems of cyber physical systems are because they rely heavily on network and data. They create a lot of problems as it is not interrelated and dependent on each other. The human system working principles can be used to solve the CPS problems.

5.2 The control systems of CPS

There are various control mechanisms inbuilt to regulate the working of the sensors and actuators in the connected network. They are Monitoring, Controlling, Diagnostics, Warning, Visualization and Self-healing.

5.2.1 Monitoring

The human system has various monitoring mechanisms in place. The body temperature is monitored by sweat glands. If it is above normal, there is sweat produced from thousands of glands to cool the body. They are all connected to the integumentary system and work in tandem with the circulatory and muscular system. There is instant feedback as sweat reduces the body temperature due to evaporation. In this case, the monitoring of body temperature involves many systems and millions of sweat glands across the body. All the sweat glands work together to produce the liquid. The skin is the sensor device and the sweat glands are the edge devices that monitor the temperature. They work independently and are indirectly connected with the brain. They are linked together by the communication systems of the human body.

5.2.2 Controlling

The brain controls the working of the human hand [13]. The control is not always directly from the brain. The tendons and ligaments, bones and arteries, muscles and skin form the model of operation. It involves four systems that work together in synchronization to enable movement of the hand. The brain controls the movement of the muscles. The modern pick and place hand robot is used for mimicking the hand movement. They work on one single operation like welding which are predefined operations. They are still being explored in robots to become general purpose like the human hands. The advantage of the human systems is that they learn and explore by themselves to understand. But machines do not have this capability to think of their own. They are fully controlled by the creator of the machine and cannot learn or act of its own. The software defines the working of the robot and it is very vulnerable if connected to the network. They can be really useful when used for dirty, dangerous and dull, repetitive tasks.

5.2.3 Diagnostics

It is to identify the cause or analytics behind the phenomenon. The headache is due to retention of fluid inside the sinus which will normally drain out by itself. This shows there is some infection in the sinus. It may also be due to stress where the brain and the eyes are tired out. Sleep and rest cures these symptom. This can be found out by an external agent called a doctor or can be self-analyzed by trial and error methods. The diagnosis happens immediately when the pain is felt. The entire body screams for rest once there is a bad headache.

The problem with CPS is the absence of pain. The wear and tear or misuse of a system cannot be found out without excessive monitoring. So now in manufacturing, fault detection, isolation and recovery is possible with the help of sensors that are deployed specifically to monitor the level of oil and general health of the machines to make remedial treatment before the machine fails. The exact location is found and recovery treatment procedure is carried out for the system. This is possible with the help of data analytics and deep learning models that can generate patterns from large amount of data.

5.2.4 Warning

The human body has two common warning symptoms that when headed gives a warning of potential disaster or danger or breakdown. They are fear and pain. They are feelings that are produced in the body to combat danger or misuse. Fear causes a flight or fight feeling to occur. Pain causes a rest and recovery procedure to be followed.

The CPS is being trained to learn this warning through the data analytics so that the decisions made can be intelligent. Artificial Intelligence algorithms have been used to generate this warning systems. They are already in place in costly and mission critical systems like nuclear reactors and space crafts to name a few.

5.2.5 Data collection

The human brain is said to have a storage capacity of 2.5 petabytes [14]. Most average humans use only 10% of the brain capacity. This small organ can store all the data present in the internet. The data collected by the brain is through the senses and stored. There are various levels of storage too. The temporary data is moved into the permanent storage area during sleep and retained. The exact method of storage of data inside the brain synapsis is not yet discovered [10].

The CPS has a constant stream of data from the different sensors connected to them. This data is very huge and it cannot be processed in the limited memory of the edge device. So most analytics happens in the cloud or data farm where artificial intelligence algorithms are used to churn out meaningful information from them. But the main disadvantage is the lack of semantics in the storage of data. Human brain automatically correlates data and finds patterns and links them together to form experiences. This correlations and linking of data has to be manually done in cyber physical systems. This has been achieved to a limited extent by programming and using machine learning algorithms. But we have a long way to go to achieve the potential of the brain.

5.2.6 Visualization

The human system has an excellent way of visualizing data stored in the brain. There are dreams that help us remember what we experienced during the day. The human memory has the ability to correlate data and find out patterns. We call it experience where training helps a person gain wisdom that is automatically applied for decision making. The method of using data for decision making has been recently made possible with the advent of the bio plausible spiking deep neural networks [11] that can do the basic tasks of classification and detection similar to a human being. The big data accumulated from the sensor networks has helped these models to learn by themselves and use the patterns for correct predictions. The industry is cautiously moving into this domain as it is nascent and not standardized. Another drawback is the unpredictability of the deep neural models when the data overfits or underfits. But the future is surely of learning systems similar to Alexa and Siri which will be integrated into all smart devices.

5.2.7 Self-healing

The ability of the human body systems to self-heal is the most intriguing aspect that has not been successfully implemented in a CPS. The fault tolerance has been researched and proactive fault recovery has been successful. But for the human body, if there is a cut in the skin, the blood clots and prevents loss of blood. The skin replaces itself with a new layer that grows and till then the coagulated blood protects the cut area. Then the covering falls off to form a new skin.

This self-healing nature cannot be replicated in a cyber physical system as the systems do not grow but are made. They follow certain principles and rules that are previously coded or generated by the intelligence algorithm.

These rules cannot form themselves. They can only define what already exists or explain what does not exist. The nanomedicine technology is trying to perform this self-healing on the human systems that need external help for recovery using targeted medicine and therapy.

6. Applications of digital twin

6.1 Agriculture

The global population is projected to surpass 9 billion people by 2050. The future climate is unpredictable and changing. Food is lost between production and consumption and the agricultural systems that grow food, fiber, feed, and biofuels need to be shrewder, robust and effective. Generating the targeted agricultural products for a smart and demanding generation will require systems that are sustainable (environmentally, economically, technologically and socially). CPS technologies will increase efficiency throughout the value chain. The proactive system must be able to monitor the growth of plants, generate proper advisory to farmers, ensure harvest at the most cost efficient time and predict climate change. Precision agriculture is a possible solution but it varies according to climate and soil type. There is a change in the weather pattern with the climate oscillating from a cold to hot weather with rains happening at one place causing flooding and drought in some other place. The ground water table is fast depleting due to misuse and over use. These cause a change in the agricultural yield. The CPS designed for agriculture should have resilient systems that can withstand changes in temperature, pressure, soil nutrients and pH to monitor and provide proper feedback.

6.2 Natural resources conservation

The natural resources are in the danger of depletion. The water cycle has been severely affected due to manmade disasters. There is a large distribution network spread over all towns for these natural resources like water and electricity. CPS technologies can provide intelligent and demand driven control and monitoring of water distribution by sensors and smart meters to enable optimal utilization of the water resources. The reuse and quality of water must also be constantly monitored. The electricity can be conserved by regulating its use and designing smart electric devices that reduce the usage by automatically turning off when not in use. The natural resources can be conserved by reducing the carbon footprint and using renewable energy sources.

6.3 Real estate

With the reduction in real estate in a populated country like India, smart and energy efficient building management systems have networked power, transportation, emergency response, and law enforcement. Smart city development must consider the energy conservation, green energy usage and energy efficient buildings. The design of buildings must be using cost efficient and renewable resources. The plan of the building can be made efficient to incorporate maximum utilization of the space for various activities.

6.4 Defense

Military and national defense has moved into deep learning faster to enable knowledge driven decision making in critical situations. Soldiers have smart wearables that monitor their health, smart food that caters to their culinary taste and contributes to the general health and well-being which is very important for a personnel posted in remote parts of the country. The defense vehicles also carry a lot of sensors and are in constant touch with the base as they take part in their critical activities.

6.5 Emergency response

Disasters are unpredictable and occur everywhere. Public safety communications are networked to improve response. All the central databases are being linked to provide immediate response. The sensor networks are being created by connecting cameras in the CCTV network to form a visual sensor network, intelligent object detection and tracking for identifying the incident, and response robotics to work in dangerous environments and inaccessible places. This will increase the awareness of emergency responders and permit optimized response through all periods of disaster events such as earthquakes, fires, floods, cyclones, accidents and terror attacks.

6.6 Energy

Clean energy from renewable energy resources such as solar, wind along with electric vehicles will become the future sources of energy. CPS technologies enable the optimization and management of resources that generate these resources, facilities that use them and users who consume these resources.

When all the devices require energy to power them, the total energy requirement explodes as the future will see an explosion in the number of sensors and actuators working to collect data. The data centers also require a lot of energy to keep them cooler and running.

6.7 Education

The education sector has jumped into the wagon of connected things. The information search has moved on to information browsing and surfing. Google has become synonymous with search. We do not store data in paper form. We have moved to digital storage. Smart classrooms have become common with smart boards and learning devices. Smart teachers have become a reality with apps developed to help children learn creatively and based on activity. The smart classroom teacher is not very far in the horizon. When all the devices in the classroom are connected, it helps in identifying problems and rectifying them faster. Accountability and transparency of the education management system increases.

6.8 Manufacturing

Industry 4.0 is the smart factory which results in data capture by integrating sensors into the industrial manufacturing machines to collect and correlate data. Cyber physical systems monitor physical processes, create a virtual twin of the physical system and help in fault detection, rectification and tolerance. They also help make decentralized decisions. All the shop floors have been connected to monitor the robotic arms that do the work in the assembly line with precision.

6.9 Transportation

Smart traffic control, analysis and monitoring is enabled by CPS systems. With the advent of driverless cars, these smart systems are used to interconnect vehicles with traffic system to prevent congestion and help travelers travel hassle free. Transportation also involves crowd control, management, energy conservation, vehicle safety and passenger safety.

6.10 Natural disaster

The unpredictable incidents are the natural disasters. They happen everywhere due to large scale misuse of nature. These disasters are preventable. But they happen due to misuse of the available resources. There is a lot of money wasted in reactive response to these disasters. A proactive response is possible with the help of cyber physical systems that monitor and stream data along with sensors that move to locations to identify the impact of the disaster.

6.11 Health care

The public and private health care industry are exploding with IoT devices that are needed for invasive and non-invasive treatment. The critical applications include surgery (laser for the cataract removal in the eye) and remote monitoring of patients health. Doctoral care is provided at the doorstep with the help of this technology. Implantables, wearables and portables are transforming the health care industry. Nano medicine has opened the door for non-invasive treatment of the human body. Robots perform surgery with utmost precision. All the body functions can be studied in a single scan. Constant monitoring and feedback with the help of wearables has increased the lifespan of an individual.

Connected smart ICUs help patients and doctors be in constant touch. The management systems in the hospital connects different departments together and provides a hassle free visit to the doctor and the patient.

7. Conclusion

The Cyber Physical systems have pervaded everywhere. It captures the data using sensors, computes and communicates alerts to the systems. The need of the hour is to look at the machines in the cyber systems as a system of systems and not as a single object/machine. The outer covering, the moving parts and connections are different layers of the same machine. When they are categorized into different sub systems then CPS will also be able to work like the human systems where the subsystems work together.

The human body has a fully connected and secure private network that can only be intruded into, to gather the vital signals of the body. It never allows external interference and the external interference causes pain and degradation of the internal system. The need in CPS is to have a fully integrated and connected private network that remotely connects with other untrusted systems only when in need. The smartness must be based on intelligent data available from analysis of the big data collected from the networked sensors. The data must be converted to experience to help with intelligent decision making. The complete process of communication, control and computation must be integrated as in the human system to help it work seamlessly. Pain and fear are two important concepts that maintain the health of the human system and is missing in the CPS.

Various examples of the human system integrated working have been discussed. The domains of interest and the possible application areas have also been highlighted. The cyber physical system must integrate the social aspect to it to become more friendly and useful. The human in the loop concept also includes the humans in CPS. But humans as an external entity will result in lack of communication in the system. Social CPS includes the humans as a subsystem to incorporate their views. Thus CPS with all the subsystems such as cyber system, social system and physical system must integrate together to seamlessly provide solutions for the problems faced by human kind.

The study of the security mechanism of the cyber physical system is a wide area that has to be explored to derive parallels to enhance the security and safety of the CPS.

References

[1] Lee J, Bagheri B, Kao H-A: A cyber-physical systems architecture for industry 4.0-based manufacturing systems, *Manuf Lett* 3:18–23, 2015. ISSN 2213-8463, https://doi.org/10.1016/j.mfglet.2014.12.001.

[2] https://flex.com/insights/live-smarter-blog/trillion-sensor-economy-coming-are-you-ready. Accessed 10 July 2019.

[3] Akbar SA, CEERI: https://www.ceeri.res.in/departments/cyber-physical-systems/. Accessed 10 July 2019.

[4] https://www.livescience.com/authors/?name=Ross%20Toro. Accessed 9 July 2019.

[5] https://opentextbc.ca/anatomyandphysiology/chapter/functions-of-the-integumentary-system/. Accessed 10 July 2019. via @pressbooks.

[6] Fass D, Gechter F: Towards a theory for bio-cyber physical systems modelling. In HCI International 2015, Aug 2015, Los Angeles, United States. LNCS 9184, LNCS—Digital Human Modeling and Applications in Health, Safety, Ergonomics and Risk Management: Human Modelling (Part I). 2015. https://doi.org/10.1007/978-3-319-21073-5_25.

[7] Xiong G, Zhu F, Liu X, et al.: Cyber-physical-social system in intelligent transportation, *IEEE/CAA J Automat Sin* 2(3):320–333, 2015.

[8] Weimer J, Ivanov R, Chen S, Roederer A, Sokolsky O, Lee I: Parameter-invariant monitor design for cyber–physical systems, *Proc IEEE* 106(1):71–92, 2018. https://doi.org/10.1109/JPROC.2017.2723847.

[9] Karthika N, Janet B: Feature pair index graph for clustering, *Int J Intell Syst*, https://doi.org/10.1515/jisys-2018-0338.

[10] Janet B, Reddy AV: Index model for image retrieval using SIFT distortion, *Int J Intell Inf Database Syst* 6(3):289–306, 2012.

[11] Janet B, Kumar RJA, Titus S: A novel and efficient classifier using spiking neural network, *J Supercomput* 1–16, 2019. https://doi.org/10.1007/s11227-019-02881-y.

[12] Thalamala RC, Janet B, Reddy AVS: A novel bio-inspired algorithm based on social spiders for improving performance and efficiency of data clustering, *Int J Intell Syst* 2018. https://doi.org/10.1515/jisys-2017-0178.

[13] http://www.eatonhand.com/hw/facts.htm. Accessed 10 July 2019.

[14] https://www.cnsnevada.com/what-is-the-memory-capacity-of-a-human-brain/.

About the authors

Dr. Janet Barnabas has served NIT, Trichy in the Department of Computer Applications for over 10 years. She has received honors include University Rank, NET for lectureship by UGC and deployed a honeypot sensor as part of National Cyber Coordination Center Project for Cyber Threat Intelligence Generation. She has published 10 papers in international journals and made 40 international conference presentations. She has set up the first of its kind Information Processing and Security Laboratory with Industry involvement. She is a champion of open source technology and activity-based learning. Her areas of specialization are Information Processing and Security, Internet of Things, Application Development and Deep Learning.

Pethuru Raj working as the Chief Architect in the Site Reliability Engineering (SRE) division, Reliance Jio Infocomm Ltd. (RJIL), Bangalore. The previous stints are in IBM Cloud Center of Excellence (CoE), Wipro Consulting Services (WCS), and Robert Bosch Corporate Research (CR). In total, I have gained more than 18 years of IT industry experience and 8 years of research experience. Finished the CSIR-sponsored PhD at Anna University, Chennai and continued with the UGC-sponsored postdoctoral research in the Department of Computer Science and Automation, Indian Institute of Science, Bangalore. Thereafter, I was granted a couple of international research fellowships (JSPS and JST) to work as a Research Scientist for 3.5 years in two leading Japanese universities. Published more than 30 research papers in peer-reviewed journals such as IEEE, ACM, Springer-Verlag, Inderscience, etc. Have authored and edited 20 books thus far and focus on some of the emerging technologies such as IoT, Cognitive Analytics, Blockchain, Digital Twin, Docker-enabled Containerization, Data Science, Microservices Architecture, fog/edge computing, Artificial intelligence (AI), etc. Have contributed 35 book chapters thus far for various technology books edited by highly acclaimed and accomplished professors and professionals.

CHAPTER TEN

Impact of cloud security in digital twin

Susila Nagarajan[a], Sruthi Anand[a], Usha Sakthivel[b]
[a]Department of Information Technology, Sri Krishna College of Engineering and Technology, Coimbatore, Tamil Nadu, India
[b]Department of Computer Science and Engineering, Raja Rajeswari College of Engineering, Bengaluru, Karnataka, India

Contents

Advances in Computers, Volume 117
ISSN 0065-2458
https://doi.org/10.1016/bs.adcom.2019.09.005

Abstract

Digital Twin is a way to virtually represent or model a physical object using the real time data. This innovation sets up a way to deal with industries and organizations to supervise their products, consequently bridging the gap between design and implementations. As the name suggests, "Digital Twin" infers that a reproduction of the product is made in order to have a nearby relationship with the live item. The procedure of computerized twin begins by gathering real time data, processed data, and operational data and performs distinctive investigation which helps in anticipating the future. This additionally enhances the customer experiences by giving a digital feel of their product. The objective behind all these is the job of gathering information and putting them in a place, i.e., the cloud which could store exorbitant data. The user experience gets enhanced by the intervention of digital twin technology which could help in the successful working of the products geographically distributed. The impact of Internet of Things and Cloud Computing lifts up the digital twin.

The information gathered from the sources can be arranged in terms of utilization and prospect to change on a timely basis. These data, as they are stored require proper coordination and a legitimate use.

Digital Twin innovation assumes incredible opportunities in the field of manufacturing, healthcare, smart cities, automobile and so on. The effect of having a digital twin for the product makes it simple for activities and recognize the blemishes, if any happened. This approach can help reduce the workload and furthermore can get trained on the virtual machine without the need of a specific training.

With the most prevailing technologies of today, like Artificial Intelligence, Machine Learning and Internet of Things more prominent approach to train and monitor products, taking care of its own execution, collaborating to different frameworks, performing self-repairs are made possible. Hence the future is getting unfolded with the emerging DIGITAL TWIN era. The massive data utilized in the field of digital twin is prone to severe security breaches. Thus digital twin technology should be handled with extreme care so as to protect the data. Hence, this chapter identifies the ways and means of collecting, organizing and storing the data in a secured cloud environment. The data is filtered according to the use and priority and pushed into the cloud. It is determined to implement an exclusive algorithm for a secured cloud which would greatly benefit the users and the providers to handle and process it effectively.

1. Introduction

Digital Twin is the digital copy of a thing of physical existence. It is an exact replica of the physical object's properties and states, including its state, position, gesture, status and motion. Digital twin is mainly used for monitoring, diagnostics and prognostics. It refers to replication of both physical and potential assets with their properties which can be used for various purposes. The concept of digital twin lies in the fact that every stage of the product development is digitized. This enables the business users having a clear notion on the working of the product in real time. When a digitized version is created, it requires plenty of data to be stored and functioned. This process can be done with the help of cloud computing technology, where any amount of massive data collected will find a place to be stored. Cloud security enforces a set of policies, rules, procedures and techniques that combine together to ensure security. They are done to protect data, customers' privacy and maintain confidentiality. They are also responsible for authorization and verification of the users and devices. Cloud security configures to satisfy the business needs by providing authentication and filtering traffic. Administration overheads are reduced since the rules can be configured and managed at single place.

2. Background study

Virtualization has gained importance for the increased demand resource, provisioning and multi-tenancy in cloud computing. Here, the prior knowledge of guest OS is unavailable. The most important considerations are IT security, integrity, availability. In virtualization, weak service oriented architecture is present which leads to key mismanagement. The main security threat emphasized here is VM hopping, where the attacker on one virtual machine gains access to the other one and delete the stored data etc. The other threats are denial of service attack, man-in-the-middle attack. The remedy for these security threats are limited resource allocation for the users in virtualization environment [1].

The Client Service Providers (CSPs) of public cloud have the full control of infrastructure and management capabilities. They have the access of direct control over systems which, sometime are considered as a threat to users who share confidential data over cloud. Cloud computing security failures could be depicted by media failures, software bugs, and malware attacks. Amazon's S3 downtime, Gmail's mass email deletions are examples which show the security failures in cloud. The threat measures can be data encryption where searchable encryption using keywords are performed. Storage auditing architecture can be done to limit auditing overhead and stricter monitoring could also be done to prevent these security attacks [2].

According to National Institute of Standards and Technology, the five most important characteristics of cloud are broad network access, on-demand service, resource pooling, rapid elasticity, expansion, measured service. The security issue in cloud still prevails without having a general purpose security mechanism. The lack of novel security in this cloud structure arises from the convergence that cloud includes such as virtualization, broadband networks, grid computing, automated systems, service orientation. The cloud computing paradigm is considered to be the great security threat which requires great research and careful measures [3].

The cloud computing architecture has been subjected to more threats for security which decreases its importance sometimes in the field of building web applications on cloud. The two security measures proposes here are multi-level architecture (MLS), atomic-level architecture (ALS). The multi-level architecture emphasizes on fine-grained access control, processing information under different classes and categories. The atomic-level paradigm performs by providing kernel space access controls such in cyber ecosystem. By providing atomic-level paradigm which gives end-to-end access for client, database, cyber applications are considered to be great security measure [4].

Cloud storage is considered to be an important feature in cloud. But the security issue for the storage of sensitive data has been found to a great threat nowadays. Hence an efficient protocol is proposed to support the batch auditing and the data operations as well. The security issue can be resolved using data fragment technique, verifiable tags. This leads to secured storage systems by providing low storage and reduced communication cost over cloud [5].

Cloud computing is slightly modifying the Internet service architecture which provides great flexibility and scalability. FRESCO, a new security application provides a script which provides monitoring logic and resolves many security key issues. Here, customization and stream processing rules detect network threats. VMware-NSX provides network system security. It proposes SDN3–7 layers of NFV software based security which decouples of underlying hardware. CNSM S- a prototype designed provides multi-tenant data centers which serves as a great measure for cloud security [6].

The storage of business data and provisioning of services are given more importance in Internet by cloud computing. The service provider could provide secured architecture for its tenants. One such structure developed here is Baseline Security which ensures multi-tenants are not having the access to attack on architecture, hosting machine. They are imposed to provide a trusted VMM platform module (TPM). The cloud boot attack can also prevented by providing security architecture as a service to cloud [7].

Data confidentiality, data integrity, data authentication, data availability are the policies to ensured for security against threats. Nowadays, some CSPs provide cyber-risk insurance which mitigates the security issue which has been provided as a great security measure to tenants. Persistent pressure on attackers of cloud, and consistent monitoring prevents or preserves the security over cloud. The cyber risk insurance is a post event compensatory mechanism which comes into play post event. So this measure is not enough to ensure security of privacy in cloud [8].

Vulnerability in cloud computing services has been increased due to many attacks by the inducers over cloud. Here, Cloud-Trust architecture is proposed which combines the IaaS and provide access controls to users. It quantifies the degree of integrity and confidentiality by estimating the high level security metrics. The proposed model consists of two Bayesian sub-networks, which provides high level of security in node classes of cloud. Security measures are not focused which adds as a drawback to the system [9].

The attack vectors of cloud include DDOS attack on cloud service providers which is great vulnerabilities. We need both proactive measures that prevent security incidents which investigates security breaches. The measures included are, for an organizational cloud user, it would also be useful to consider the different rules in which the cloud services are hosted. Those cases must be identified and resourced by other means. It would purely vary on the deployed architecture [10].

Cloud computing security has become a limitation in its development. Physical security, network security, host security, abstract resource securities are some of the security concerns to be considered. At the general level cloud security, cloud system should be able to identify natural disasters and at enhanced level, system should be able to identify serious security threats. In feedback assessment, five attack paths have been designed using vulnerability scanning and external penetration testing. Comprehensive feedback assessment method has been used to check the cloud assessment indicator for system's validity, reasonability [11].

The cloud assurance for small firms is being provided by frameworks which have proven limitations. The proposed PDCA approach can be employed in small enterprises comprising IT risks. cloud-adapted risk management framework (CARMF) addresses risk management security issues. In order to reduce security overheads standards such as ISO/IEC 19086 can be introduced. These methods enhance security over cloud for business firms [12].

The adoption of cloud computing has been a discussion over the years. In this paper, a framework has been developed to secure cloud data. There are layers for security which includes firewall and authorized access, infrastructure management, intrusion detection and encryption. CCAF with

BPMN together provides a better performance in evaluating the security aspects. Thus CCAF is found to be a better approach for security concerns [13].

Information security risks are one of the major driving factors to be considered in security of cloud. This paper emphasizes on the role of organizations to quantify the residual risks below the threshold of the acceptable level of risk. Risk management framework (RMF) applied to cloud ecosystem can be used to address the risk factors in security. These measures can be used in organization to monitor security attacks [14].

The Cloud Certified Security Professional (CCSP) curriculum is designed to validate information security professionals in terms of their competency in their relevant field. The CCSP is most appropriate for the projects that involve procuring, securing and managing cloud environments or managing the purchased cloud services. These measures of training the professionals comprehensively cover the risks and security concerns associated with cloud computing [15].

3. Digital twin

Digital twins are virtual representations of the physical products, assets and processes which would greatly help in analyzing and optimizing the performance of the physical entity in order to achieve improved business outcomes.

Digital twins consist of three components:
- A data Model
- A set of Analytics or Algorithms and
- Knowledge.

3.1 Data model

A prototype depicting the components, structure and its modules guides the user to work in a virtual space with a feel of working directly with the physical product.

3.2 Analytics

Using suitable machine learning techniques helps us to analyze the performance of the system so as to predict the futuristic behavior.

3.3 Knowledge base

The reports generated through analytics along with the subject expertise, historical data, and industry best practices are stored in the knowledge base for future references and continuous learning.

4. Advantages of digital twin

- Reduced risk.
- Improved production.
- Increased reliability and availability.
- Low maintenance cost.
- Faster time-to-value.

5. Characteristics of digital twin

The two most important characteristics of digital twin are:
- Relationship between physical and its virtual entity.
- Used for generating Real Time Data using Sensors.

In addition, it integrates various emerging technologies like Machine learning, Artificial Intelligence and IoT. By combining all these smart technologies, it updates and modifies itself with respect to changes in physical environment. Digital twin gains its input from various sources like sensors which provide real time information. The Digital twin consists of three major parts: Physical product, Virtual product, Connection between Physical and Virtual product. The concept of digital twin is further divided into three major types: digital twin prototype (DTP), digital twin instance (DTI), and digital twin aggregate (DTA) as shown in Fig. 1.

- The DTP include the steps of designing, analyzing and processing techniques to understand physical entities.
- The DTI is the digital twin of every individual instance of a physical product in the process of manufacturing.
- The DTA is the aggregation of DTIs. The physical entity is analyzed using the data and information collected.

The main characteristics of digital twin technology are Connectivity, Homogenization, Smart and Reprogrammable, Digital Traces.

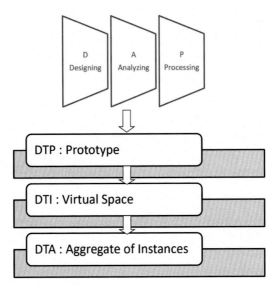

Fig. 1 Components of digital twin.

5.1 Connectivity

Internet of Things is one of the prevailing technologies in the recent time. The Digital Twin technology has similarities with IoT. Most importantly, this technology enables connectivity between physical component and its digital counterpart. This connectivity acts as a base for the Digital twin technology. Its works by collecting data through sensors fixed in physical product and integrate these data for usage in various growing technologies. It plays a vital role in establishing connectivity between various customers and their products.

5.2 Homogenization

Digital technology is the enabler of homogenization of data. It is used to create visual representation of data as information can be changed into digital without change in its properties. Nevertheless, it's a process of decoupling data from its original physical form. Homogenization and decoupling lead to user experience convergence. Due to digitalization a single entity can have multiple forms. Howsoever, Digital twin technology allows its users to transmit information about a physical object to various sectors on an industry. For example, in conventional methods of industry maintenance, only the authorized personnel's have the privileges to monitor the working conditions of the machines in the factory floor. Digital twin supports everyone connected with that sector massively by enabling easy monitoring in not just a single plant but for numerous factories.

5.3 Reprogrammable and smart

The unique feature in Digital Twin technology supports modification in the physical and the virtual instances by suitable reprogramming. When changes are made in physical entity the alterations in the digital twin can be automated and vice versa. These changes are generally made by manipulating the data received through sensors.

For example, in automotive sector, when an additional feature is required in an engine for an increased performance, it can be reflected through its digital twin without the need for physical intervention.

5.4 Digital traces

Digital Traces are generally used to identify the cause for the malfunctioning of the machine. Let's say, for example, when a machine fails in its proper functioning, it gets back to the previous state or stop at that particular point which leads to a major production loss. In order to overcome this point of failure and to predict the same in early stage the Digital Twin is reliable.

5.5 Modularity

To improve the performance of the product used, modularity plays a vital role. Using this feature, the changes to be made can be easily tracked and modified. The various metrics enable us to visualize how the entire functionality is administered. When a machine or a product fails, the fault will be identified by bringing the machine to the initial state. The impact of the digital twin technology helps us monitor the functionalities of the product at once the point of failure is noticed thereby reducing the production loss.

5.6 Combination of cloud and digital twin

The digital twin in the cloud collects data from its different sources depending on the historic, present data from the labs and also from feedstock and energy pricing. The data are analyzed and monitored by the experts and optimized to improve performance through highly efficient cloud based system.

High-level security concerns like access to unauthorized data, weak access controls, susceptibility to attacks, and availability disruptions affect the cloud systems. Cloud security architecture becomes successful only if the defensive mechanisms adopted are successful. An efficient system should solve all the issues that focus on security management.

6. Cloud security techniques

6.1 Data encryption

Encryption finds to be a great approach regarding data security. Encryption of data before sending it to cloud is a necessary part in cloud. Only the data owners can allow privileges to access that data. The file being sent to cloud will be sent as an encrypted file wherein the service provider further enhances the encryption with respect to their standards. This process is known as multistage encryption. To provide better encryption on data, multiple algorithms are employed. This is done to avoid unauthorized access from other users, making the data unavailable to other user's. The virtual instance created as a digital twin for any physical prototype requires this multistage encryption in order to avoid data loss and to protect the physical product.

6.2 Hashing

Hashing function generates a hash value for protecting the data being stored in the cloud. This hash calculation can be used for data integrity but it is difficult to decrypt while the hash value is the only known key. The keywords and private keys can be maintained proper before the storing the data in the cloud. RSA based data integrity checks identity based cryptography and RSA Signature. Credentials of individuals or attributed based policies are better utilized to differentiate the unauthorized users. Permission as a service can be employed to instruct the user about the specific location of the accessible data. Fine grained Access control mechanism provides an efficient way to allow the owners to delegate highly computational tasks to the cloud servers without data leaks. In order to accomplish this data driven framework can be designed for data sharing. Network based intrusion prevention system can be adopted for real time detection. RSA based storage security solves the problem of remote data security.

7. Cloud security in digital twin

The most popular cloud service provider such as Azure has created a digital twin called Azure Digital Twin which gives more compatible security of digital twin in the cloud. The basic procedure or process undertaken to

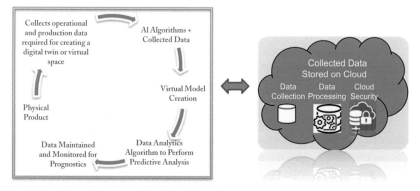

Fig. 2 Work flow diagram—Cloud and digital twin.

provide cloud security can be provided while collaborating digital twin in the cloud is shown in Fig. 2. It includes monitoring for prognostics, and collection of operational data on virtual space, which results in virtual model. And further algorithms are applied to store the data secured on cloud.

8. Impact of digital twin in aircraft

For example, let us see how digital twin plays its role in the field of Aircraft and impact it gives in prognostic analysis.

The Aerospace Industry has been using the Digital Twin technology for many years. After the launch of Apollo 13 on April 1970 [16], no one could have predicted the explosion of the oxygen tanks, early into the mission. The issue was identified just 200,000 miles away. NASA ended up with a swift rescue mission by operating Apollo 13's digital twin model on earth which allowed engineers to find out the best possible solutions through the virtual instance and they successfully prevented the explosion. From there on, the Digital Twin technology started emerging gradually. In the Aerospace Industry, this Digital Twin technology is mostly used to develop digital models of commercial aircraft and air force jets in order to find ways to improve and achieve optimal performance of the physical machine. A fully functional digital twin offers comprehensive and predictive analytics. An aircraft incorporated with the digital twin technology would be able to help in prognostic failure based on the accumulations of previous data. It would allow the aircraft technicians to examine the issue before any risk arises and reach a conclusion that whether it required a full re-testing of an aircraft's airframes, testing its engine or carrying out any further safety checks to guarantee the safety of the individuals on board. It enables idea generation and its testing, even before the actual manufacturing begins. This technology not only helps the aircraft operate effectively and reduce testing expenses but also to manage and fix failures in the system even when they aren't within the physical proximity.

A digital twin makes it possible to predict the life of the physical asset with a high level of accuracy. For example, in a digital twin of the landing gear, sensors are placed on typical failure points like hydraulic pressure and brake temperature. This facilitates the collection of real time data from these points. The processed data helps in the early prediction of landing gear malfunctions and helps in determining the landing gear's remaining life cycle. NASA is already harnessing the power of the digital twin to craft flawless blueprints, roadmaps, and next-generation vehicles and aircraft. Simulations that are created via digital twinning allow direct, real time monitoring of the health and structure of plane parts, thereby reducing the need for manual examination of aircraft engines. For instance, if the digital twin data is indicating that the overall life of an engine is going to end in 2 weeks time, specialists can release orders stating not to allow the vehicle to leave the airport. One of the biggest advantages is that, by having full transparency on the components of an aircraft, engineers are able to see which parts are experiencing weight strain.

For instance, an aircraft that carries cargo often has a lot of limitations in terms of how many goods it is able to carry. As it is virtually impossible to make an accurate estimate of exactly how many goods can be carried, pilots must fly with less cargo to ensure safety.

However, if they would start twinning aircraft and work on simulating flights, they could assess how many goods the plane is able to carry. But, there are a lot of factors that need to be taken into consideration when judging the overall weight an aircraft can carry. Some of those factors include weather conditions and air pressure that differs with height. Digital Twin cannot exactly simulate weather, but it can gather data with every single flight an aircraft experiences and then take the gained information to determine the requirements, which are then used in the equation used for determining weight limits. The data of the digital twin can be fully trusted only by conducting many experiments flights on an aircraft. Since a digital twin is linked and updated on every event, it serves as a manual with all system data to assist airlines to function error-free considerably.

Thus, digital twin is the sort of technology that has enabled the aerospace industry to evolve and to make modifications to improve effectiveness and to bring about numerous advances.

9. Use of Skyhigh

The users of Skyhigh get the privileges to monitor the user activity, segregate the sensitive data and check the breaches in IT Cloud services to propose suitable DLP Policies. The sky High security provides the user with rich features enforcing security such as data protection, segregation of tokens from the substantiate data and access control. The different types of attack like insider attack, attacks from authorized users and compromised accounts are identified and actions are taken.

9.1 Features and working of Skyhigh

The main features of Skyhigh include Cloud Registry, AI-Driven Activity Mapper, User Behavior Analytics, Access Analytics, Security Configuration Audit, Insider Threat Detection, Multi-Tier Response, Collaboration Control, Autonomous Remediation and Encryption.

Skyhigh Security ensures the protection of data traces left over in the mobile devices and desktops, when the user access the cloud services through the enterprise networks. By this way it also eliminates the need of VPN and backhaul. Hence the mobile to cloud connectivity can be enhanced, thereby increasing the capabilities to access secured cloud services. The threats are detected by performing several analytics in different types if data like the users, usage, nature of the application etc. With these enhanced features, Skyhigh can provide easy installation and use by simple three step configuration, flexible deployment and promising delivery.

The Skyhigh security identifies the threats of multi user login and identifies the anomalies in it in order to prevent security breach. It analyzes the login patterns of a single user to check whether the same user login us traced in multiple locations. Adaptive authentication technique is employed as a reformative control which brings in the multifactor authentication such as malicious login. With this technique an extra layer of security is imposed in order to prevent the user login falling into the category of compromised account.

10. Applications of digital twin technology

10.1 Monitoring

Combination of the data received from the sensors of physical entity with digital twin provides high level of monitoring. One of the major examples includes 3D visualization of car. Using this technology every part of the car can be easily viewed and monitored.

10.2 Training

As digital twin provides the exact replica of a physical entity it can be used for easy learning since it provides a visual picture of the events. It educates the users with a clear and accurate knowledge of the task undertaken.

10.3 Strategy

This technology can be used in the process of optimization without downtime. The test cases are applied to the virtual entity and on success of the test on virtual product, it allow us to implement the process on the physical entity.

11. Conclusion

Technology finds its way to an interesting and burden less era. Digital Twin benefits the industries by enhancing the production and handling efficient supply chains. The twin created in the virtual environment monitors and keeps track of the events that occur in the real product and through the records created by the past, present and future data it helps in predictive analysis. With the combined effort of our machine learning and AI algorithms, digital twin impacts the modern world to have increased customer service and trust that reduces the failure which in turn plays a role in the life of mankind. This technology reduces the risk of failure by early prediction with the twin in the virtual space, thereby updating it in the physical device. This can be done with the help of Digital twin lending out data that is received through its sensors for the purpose of predictive analysis. When operational and production data has been received, it is stored onto the cloud for the further process. The data on the cloud has to be monitored and secured to enable trustworthiness of the provider. Hence, the role of digital twin has a major influence on the business operations that provide a greater advancement toward the technological changes and binds in with machine learning, AI, etc. Thus various techniques and algorithms have been imposed to provide security in cloud based applications. The cloud computing service provider and the customer should be aware about protection of cloud from all the external threats or attacks, so there will be a strong and mutual interpretation by both of the customer provider. One important feature of digital twin is to simulate the data analyze and predict events and situations, which is done through the virtual space or the instance. The analysis is done through the historical data and machine learning algorithms which help in the process of predictive analysis.

References

[1] Tsai H-Y, Siebenhaar M, Miede A, Huang Y-L, Steinmetz R: Threat as a service? Virtualization impact on cloud security, *IT Pro* 14:32–37, 2012. *IEEE Computer Society*.

[2] Ren K, Wang C, Wang Q: Security challenges for public cloud, *IEEE Internet Comput* 16:69–73, 2012. IEEE Computer Society.

[3] Mell P: What's special about cloud security? *IT Prof* 14:6–8, 2012. IEEE Computer Society.

[4] Brown A, Apple B, Michael JB, Schumann M: Atomic-level security for web applications in a cloud environment, *Computer* 45(12):80–83, 2012. IEEE Computer Society.

[5] Ni J, Yu Y, Mu Y, Xia Q: On the security of an efficient dynamic auditing protocol in cloud storage, *IEEE Trans Parallel Distrib Syst* 25(10):2760–2761, 2014.

[6] Chen Z, Dong W, Li H, Zhang P, Chen X, Cao J: Collaborative network security in multi-tenant data centers for cloud computing, *Tsinghua Sci Technol* 19(1):82–94, 2014.

[7] Varadharajan V, Tupakula U: Security as a service model for cloud, *IEEE Trans Netw Serv Manag* 11(1):60–76, 2014.

[8] Tari Z: Security and privacy in cloud computing, *IEEE Cloud Computing* 54–57, 2014.

[9] Gonzales D, Kaplan JM, Saltzman E, Winkelman Z, Woods D: Cloud-trust—A security assessment model for infrastructure for clouds, *IEEE Cloud Computing* 4(3):523–536, 2017. *IEEE Computer Society*.

[10] Juliadotter NV, Choo K-KR: Cloud-attack-risk assessment taxonomy, *IEEE Cloud Comput* 2(1):14–20, 2015. IEEE Computer Society.

[11] Chen X, Chen C, Tao Y, Hu J: A cloud security assessment system based on classifying and grading, *IEEE Trans Cloud Comput* 2(2):58–67, 2015.

[12] Luna J, Suri N, Iorga DM, Karmel A: Leveraging the potential of cloud security service-level agreements through standards, *IEEE Cloud Comput* 2:32–40, 2015. Published by the IEEE Computer Society.

[13] Chang V, Ramachandran M: Towards achieving data security with cloud computing adoption framework, *IEEE Trans Serv Comput* 9(1):138–151, 2016.

[14] Iorga M, Karmel A: Managing risk in a cloud ecosystem, *IEEE Trans Serv Comput* 2:51–57, 2015.

[15] Gordon A: The hybrid cloud security professional, *IEEE Trans Serv Comput* 3:82–85, 2016.

[16] https://www.space.com/29078-how-apollo-13-moon-accident-worked-infographic.html.

About the authors

Dr. Susila Nagarajan is currently working as Professor and Head in the Department of Information and Technology at Sri Krishna College of Engineering and Technology. She has a total experience of 18 years. She has graduated from Sathyabama University with 9th rank and also completed her Ph.D., in the field of Cloud Computing thereafter. She has published papers in various National and International Journals. Guided a project titled **"INTELLBOT"** for the 2000–2004 batch students, which won the **FIRST PRIZE** in **Indian National Academy of Engineering (INAE)**, Delhi. Guided a project titled **"Voice Based Automated Wheel Chair for Handicapped"** for the 2009 batch students, which was awarded a grant of Rs.5000 from the **"Tamil Nadu State Council for Science and Technology"**. Received **"Best Faculty Award"** for the year 2015 at Sri Krishna College of Engineering and Technology. Organized various Workshops, Conferences, Power Seminars and many more, she is also the organizer of the first edition of the TeDX Event. She is also a reviewer for the Journal of Cloud Computing: Advances, Systems and Applications, ISSN: 2192-113X (Online), Springer Open, Editorial Review Member for International Journal of Information Security and Privacy, Review Member for Expert Systems, Wiley. She has been an Academic Jury Member for Youth Talk—ICT Academy and Session Chair for National and International Conferences. She has also published a book chapter titled "Impact of Cloud of Clouds in Enterprises Applications" in the Book Novel Practices and Trends in Grid and Cloud Computing published by IGI Global.

Sruthi Anand is working as an Assistant Professor in the Department of Information Technology at Sri Krishna College of Engineering and Technology. She is young and dynamic with 3 years of experience in teaching. Having graduated from SRM Institute of Science and Technology, her interests include Databases, Computer Networks and Cloud Computing. She has published papers in various Scopus Indexed Journals. She has received the "Young Achieving Faculty Award" during the year 2018 and 2019 at Sri Krishna College of Engineering and Technology.

Usha Sakthivel is currently working as a Professor and Head, CSE, RRCE with an experience of 21 years. Graduated from Manonmanium Sundaranar University, in Computer Science and Engineering during the year 1998. She obtained her Master degree in Computer Science and Engineering and Ph.D., degree from Sathyabama University in the area of Mobile Ad Hoc Networks in the year 2013. She has 54 publications in International and National conferences, 22 publication in National Journal and International Journals in the area of Mobile Ad hoc Networks and wireless security. Most of the publications are having impact factor citied in SCI, Google Scholar, Scopus (h index and i10index), Microsoft etc. Received fund from AICTE under NCP Scheme, MODROBS, TGS and SERB(DST). Received best teacher award from Lions club in the year 2010 and 2012. Received best paper award in many conferences. Developed Centre of Excellence lab in IoT with industry collaboration. Organized many conferences, FDPs and Technical Talks. Associated with ISTE, CSI, IEEE, IAENG, IDES and IACSIT. Reviewed papers in IJCs and CiiT journals. Acted as a TPC member in MIRA'14 IoTBDS '17 and IoTBDS'18 Portugal. Chaired sessions in FCS'14, ICISC'13 and ICCCT'15. ICCCT'17 and IoTBDS'18. Local chapter Active SPOC for NPTEL, College website co ordinator and NBA co ordinator at college level.

CHAPTER ELEVEN

Digital twin in consumer choice modeling

D. Sudaroli Vijayakumar
PES University, Bengaluru, India

Contents

Abstract

"Digital twin" more often perceived as a twin terminology along with industry virtualization of physical assets. The usage of digital twin on physical asset is well known, such as to predict when the individual parts of a machine must be replaced. However, digital twin technology in non-physical modeling is a vibrant research area. One area where digital twin can be effective is predicting the customer's needs. Most businesses to predict the customer's needs uses risk analysis and profitability assessment which holds its own pitfalls. One of the major downfalls arises during the analysis on historical data is the time consumed.

Time is one of the crucial factors that determines the profit a company makes, holding of customers, satisfying the customers' needs at the right time, fails because of the

Advances in Computers, Volume 117
ISSN 0065-2458
https://doi.org/10.1016/bs.adcom.2019.09.010

© 2020 Elsevier Inc.
All rights reserved.

265

static behavior. This can be made more effective by enforcing Digital twin to track the customer behavior dynamically such as the products they consume, their satisfaction. So instead of relying on the historical data, the data for digital twin will be from CRMs, logs, order processing info etc. Right product at right time can be achieved by creating suitable machine learning models on this dynamic dataset and this trained model are held in the digital twin, which runs them in real time. For achieving this approach, a specific technology called Tarantool Data Grid is very useful. In this chapter, we will explore how this technology can be used to create consumer choice modeling using Digital Twin with suitable use cases.

1. Introduction

Digitalization is transforming everything from people to organizations, the way we live and work. The era of digitalization demands a solution that is practically feasible and achievable. The pragmatic solution to any problem can be obtained with the aid of virtualization technology called "Digital Twin." Digital twin lays the foundation for intelligent applications with its versatile nature providing an excellent opportunity for the human population to scrutinize the abnormalities and identify the best solutions. The digital twin is a technology that allows us to design, test and build systems for everything in the physical world [1]. Data-driven product design has emerged with Industry 4.0 and has given exceptional results in building, managing and maintaining a physical asset. The outcome of customer-centric industries may be products or services. Modeling the consumer choice is comparatively tedious as the business model is linked to customer satisfaction, thus requires creating a behavioral digital twin. Creation of behavioral digital twin requires not only a huge collection of data but intelligent analysis to understand the consumer characteristics. In recent days, there are some terminologies commonly used to do big data analysis is machine learning. However, the similarities and differences between big data, machine learning, artificial intelligence and digital twin is essential to merge them appropriately and build a highly accurate decisive system. This chapter reviews the concepts of big data, machine learning and digital twin in consumer choice modeling considering retail. On this basis, they are compared from different aspects.

The main contributions of this chapter include:

1. Concepts of Big data, machine learning and artificial intelligence are reviewed along with the understandably of data applications in consumer retail.

2. The differences between these terminologies is discussed and how the digital twin, big data, machine learning and artificial intelligence can be joined to promote smart consumer choice modeling.

3. Considering retail as an example, how these four technologies can be clubbed together to produce a behavioral digital twin is explored by understanding the drawbacks occurring in conventional retail.

The rest of the chapter is organized as follows: The concepts of big data and machine learning in retail are reviewed in Section 2, followed by the digital twin in Section 3. The drawbacks of traditional retail are presented in Section 4. In Section 5, the methodology to produce a behavioral digital twin along with the Tarantool is discussed. Finally, conclusions are drawn in Section 6.

2. Big data, machine learning and artificial intelligence in retail

Data is becoming the important assets of human society, and big data era has come [2] because of the increment in the number of physical devices and their connectivity. Validating such huge data in terms of cost and time is the much-needed solution for every prevailing problem. Almost all industries for decision making is completely relying on data analysis and not experience. Retail is nothing away from this, and these hot technologies are trying to discuss the problems faced in conventional retail effectively.

2.1 Concept of big data, machine learning and artificial intelligence

Though all the three terminologies are used frequently nowadays, the concept underlying each of the terminologies is different as well as its function. Irrespective of huge popularity, the term Big data still do not have a unified definition. Any data which possesses 4V's can be considered as Big data. The data which is huge in volume, and it is a combination of structured, semi-structured and unstructured data, and the rate at which data is getting generated is faster and its significance is not the only volume but the value it retains within itself. An appropriate example to understand big data in a clear way is an answer to this simple question: How many users are accessing Google in 1 min? Stats says that there are 4.1 million users. These 4.1 million users' searches, clicks are all data and thus it shows clearly that these types of data are Big Data and the data scale is very large, ranging from several PB (1000 TB) to ZB (a billion TB) [3]. Furthermore, the characteristics of big data are extended to 10Vs, i.e., *Volume, Variety, Velocity, Value, Veracity, Vision, Volatility, Verification, Validation, and Variability* [4].

As specified earlier, the significance of Big data is not in terms of volume but deriving value out of it. To extract value from the massive data requires the usage of powerful algorithms. Machine learning is a technology that tries to derive value from the huge data. The machine learning model is completely different from traditional programming (Fig. 1).

The input to a computing environment will be the combination of the input and output data which in turn tries to create a machine learning model/program to answer various analytical questions. Going back to the same google example, the number of users who accessed Amazon by clicking from google. This can only be answered if we have fed in the entire set of all types of data generated by the Google users. Depending on the type of answer we are requesting, appropriate algorithms must be used to create the models. This model, in turn, will serve as an agent who can provide value to the acquired big data.

Artificial intelligence and machine learning are two terminologies that are often used interchangeably irrespective of its major differences. Artificial intelligence simply can be defined as making a man-made object to think and thus it always demands not the only solution as happens in machine learning but an optimal solution thus leading to success, not accuracy. These three technologies used collaboratively have made tremendous differences in the way retail businesses were dealt with. However, along with the above technologies if the digital twin is integrated can make massive changes in consumer choice modeling.

2.2 Data sources for retail

In retail, big data refer to the data generated from the purchase retail life-cycle, such as innovation, accelerated development, maturity, and growth [5], which are also featured with 4Vs. Retail data are generally collected from the following aspects:

1. Customer data that can be collected at the point of sales include types of items sold, prices of items sold, total sales for the day, total sales by

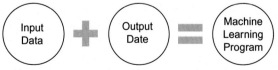

Fig. 1 Machine learning approach.

category and customer data that will help in identifying the best customers, best product, worst product and the highest sale hit days. In the case of online retailing, the type of data can be obtained from the customer's order history.

2. Management data from order management systems to enhance the promotional strategies in turn can increase sales.

3. Internet data including the users accessing the e-commerce platforms and social networking platforms to effectively perform sentiment analysis and recommend products according to their mood and behavior.

These retail raw data can be barely useful. It must be pre-processed to make it suitable to perform retail analytics. Since the data is collected from multiple sources, it typically results in heterogeneous, multiscale data. This data should be cleaned to identify missing values, collinearity features, etc. The valuable knowledge is extracted from many dynamic and fuzzy data, enabling online retailers to deepen their understandings of various stages of the retail lifecycle. Therefore, online retailers will make more rational, responsive, and informed decisions and enhance their competitiveness. One of the major advantages of using Big data platforms is allowing the complete data for analysis which was not a feasible option during the earlier stages [6]. With this widened technology, retailers can apply big data to include price optimization, customer micro-segmentation, marketing, inventory management, customer sentiment analysis, and in-store behavior analysis [7].

2.3 The applications of AI, ML and big data in retail

There is a paradigm shift in the way retail business is handled. Product-centric the approach of selling products to consumers has become an alien approach nowadays and the consumer-centric approach of maintaining relationships with consumers has become the mantra for business in retail. Omni channel shopping experience no more exists, and the consumers have numerous choices in selecting the shop for purchase, payment and delivery options. Digital transformation is fast pacing and the most vital entity of retail business that is the consumers has undergone a makeover and emerged as a digitally legitimize tenet. In the retail landscape, the revolution in moving to digital space was initialized by Amazon and eBay and the success is mainly because of their ability to derive perceptive and prosecutable insight from the data. The intersection of three technologies Artificial Intelligence, machine learning and big data gave the retailers tangible acumen from multiple sources data. Some of the areas where these three technologies have shown tremendous growth in sales for online retailers like Amazon are as follows:

- One of the primary factors that ensure significant upsell in retail is forecasting demand and taking appropriate supply decisions. Recommendation engines help online retailers in achieving this [8]. Current recommendation engine generally investigates the purchasing pattern of buying customer as well as the similar consumers. Machine learning algorithms like collaborative filtering, content-based filtering, clustering, and categorization are generally used to make customizable recommendations.
- Inventory planning earlier used to be a headache with several trial and error took a twist with machine learning based data analysis. The root cause for any problem is accurately identified with the aid of learning algorithms. Incomplete data, human bias, and guesswork are eliminated providing more space for doing better segmentation among customers. Artificial intelligence is used in many areas where the human process can be automated to provide better promotions, assortments, and supply chain thus offering more personal and convenient shopping experiences.
- Same day delivery from online retailers eased customers experience in an online purchase and this combination of dynamic routing and constant performance optimization bodes well for retailers seeking to optimize efficiency and customer service.
- Another interesting application introduced by online retailers in shopping assistants that can work several percent's efficient than that of the human sales assistants. Machine learning integrated all the product tracking information to provide customers personalized shopping experience.
- One of the major issues that the retailers face is giving only promotions that are very likely to deliver a satisfactory return on Investment. Since the machine learning algorithm already simulates the potential outcomes of promotion, potentially risky promotions are eliminated and recommend only the promotions that can boost sales and profits.

The retailers gained better efficiency and productivity with the increasing use of machine learning, AI and big data, however, there is a real need for automation to meaningfully connect with consumers so that there is returned customers who experience complete shopping only on their websites. This is the focus of integrating digital twin on top of these three technologies.

3. Digital twin in retail

Online retailing has come a long way and these e-commerce brands continuously working to decrease operating costs and to build a better relationship with customers. Digital twin is one technology that serves as a bridge between the physical and virtual world providing the e-commerce brands for better Return on Investment.

3.1 The concept of digital twin

It's a digital representation of a real-world object, product or asset. The concept of a digital twin has been around since 2004, however, it gained strength with disruptive technologies like the internet of things (IoT) and cloud that drastically brought down storage and processing cost. The term Digital Twin was first published by National Aeronautics and space administration (NASA) in 2010. Replicating physical systems with the assistance of best available physical models, sensor and historical data [9].

The typical workflow of a digital twin is depicted through the following diagram (Fig. 2).

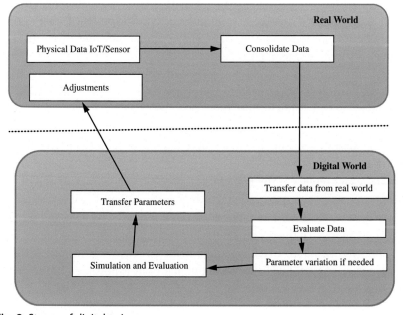

Fig. 2 Stages of digital twin.

Relevant data from the real world is collected through various IoT sensors continuously and transferred in real time to the virtual replica of the system called the digital twin. Data from real-world are sent to virtual models and then simulation and validation is performed, and that output of the virtual world data is again fed back into the physical world if any improvement is required.

3.2 Digital twin in consumer choice modeling

Digital twin adoption in manufacturing greatly reduced the length of product development lifecycle without compromising on the accuracy, stability and quality of the product. Digital twin in manufacturing is the digital simulation of physical machine through which several what if scenarios of the product is tested. Engineers enjoy the benefit of testing and updating of digital models thoroughly and thus the scope for trials and errors is minimal. With the same context if digital twin is used in consumer choice modeling, customer experience can be greatly improved. Success in retail highly depends on the substantial customer base. Growing and maintaining this customer base is essential and digital twin can play a significant role in enlarging customer service. Regular online activities in the form of personas, social media and store visits of the consumers are recorded to create a digital twin model. This in turn can create a unique customer experience such as providing cosmetics based on their digital twin model. In order to fit the consumer's pattern of interest, the life of the supply chain in retail would be shortened. Irrespective of tremendous shift of shopping experience from brick and mortar to digitalized platform, the craze to physically touch and feel the product and buy is something that can never be compromised from consumer's point of view. Digital twin can be adopted under various phases when it comes to consumer choice modeling in retail. Some of them are as follows.

The exact replica of the real physical product can be maintained on the internet to make necessary updating as it goes through physically. One of the major drawbacks that is faced in retail production is the wastage and this can be greatly reduced if the product is tracked throughout the entire life cycle of the supply chain. This tracking can be made effective and fruitful with the abetment of Digital Twin. Successful digital twin is possible only if simultaneous metric measurement can be carried out unlike analytical models. A simulation model that can actively measure the disruption duration and performance impact will make up a successful digital twin. Digital twin of the supply chain mirrors the transportation, inventory, demand and capacity of the physical supply chain that can be used for real time decisions on the product. This adoption of digital twin in supply chain can make huge difference in daily business operations as this is a balancing act between logistical variations and customer requirements. This ensures identification of low running processes, delays and root causes that optimizes the supply chain process.

Supply chain digital twin enforces right products at right time, promoting and achieving higher revenue is the next desired parameter. Recommendation engines created using machine learning algorithms do contribute toward revenue generation, however, the future of retail is insight driven relevancy. Recommendation engines primarily created by observing the consumer's transactional data irritates them by its untimely notifications. This directly contributes on the targeted sales thus requiring promotion of sales according to the individual customer behavior. Today's consumer demand relevancy and that can be made possible by combining both the transactional and non-transactional data. Applying these insights retailers can communicate with the consumers when it's really required for the consumers to buy. With the transactional data, retailers will know the brand of the coffee a consumer is purchasing as well as the frequency. By observing the usage of consumers social media retailers can very well understand if the recommendation of coffee brand is required or not. If the consumer is away for vacation, then such type of recommendations can be avoided. With the aid of transactional and non-transactional data, relevant and timely suggestions like their favorite coffee brand is available at the local stores which are conveniently located on their route home from work and ordering includes complimentary gifts. This methodology of understanding the consumer more personally and giving relevant information by observing both the transactional and non-transactional data can give additional profitable revenue.

In consumer choice modeling, customer satisfaction is the key parameter. Another factor that requires digital twin adoption is the way risk and compliance activities are managed. Addressing all risk and compliance activities and initiating improvement activities helps to satisfy the customers in a better way. With the aid of Digital twin, this risk and compliance activities can be enhanced in identifying the best practices among the diverse regulations. Digital twin abidance with the above-mentioned factors will enhance the retail business to a greater extent, however, the major hurdle for retailers and distributors is logistics. Manually checking and listing the items that must be redistributed can be error prone. This can be effectively addressed by combining transactional data, social media contents as well as the data from IoT enabled devices like smart phones, home appliances and cars it becomes possible to create digital twin for consumers. This can provide consumers a relevancy-based suggestion without irritating with multiple recommendations.

4. Drawbacks in traditional retail

The offline retail is facing major problem because of the time constraint among the busiest population today and they prefer buying at their time leisurely and the products are delivered at doorstep. Except the higher investment in place and buildings one cannot replace the joy the consumer experiences with physical shopping. In terms of managing the supply chain, order processing both the online and offline retail needs changes in the way it is been handled. Consumers selecting online or offline retail is a choice. But as far as the retailers are concerned, the more pressing challenge is to create an omnichannel strategy of consumers which definitely demands a technology that can help retailers to exhibit them better either online or offline.

5. Technologies to build digital twin

Digital twin implementation is generally carried out in an organization that is matured as well as flexible enough to deliver the promised value. A digital twin adoption cannot happen to serve a single purpose. Taking our retail itself as an example the entire cycle should be created as model not concentrating only the logistics. If the model is meant to be built for a single purpose, it will become obsolete over time. So, building a dynamic digital twin requires the maturity and flexibility.

The process of building a digital twin is illustrated in Fig. 3.

5.1 Design

The driving force behind building a successful digital twin is the type of information that is required across the entire life cycle may it be supply chain

Fig. 3 Typical digital twin workflow.

digital twin, risk governance twin etc. In the design phase of digital twin, we need to identify the type of information required, where it is going to be stored and how it is going to be used.

Along with the required data, design phase should also have adequate information about the technology that can integrate the physical and virtual world. Getting very closely with our objective of creating a digital twin for consumer choice modeling, the information required to build the supply chain twin can be collected by attaching IoT sensors to the assets and equipment. Supply chain should be capable enough to handle any potential planning disaster. To enrich the current planning process, customer demand, manufacturing and logistics information is required. The type of information that can be collected includes the ones which is mentioned under [10]

- Information about the customers both in the form of structured and unstructured. Trending products, events that affect the demand and delivery of a particular product, weather changes that has huge impact on the sales of a particular product will help us to understand the customer demand in a more meaningful manner and these can be source of information for the twin.

- Manufacturing process of a product can enrich the supply chain process by giving answers to various queries like the maximum capacity, current production, adjustment to current plans, etc.

- Digital twins of your manufacturing facility, warehouse operations, and transportation processes provide glimpses of your existing and projected inventory—as well as where to find it. And with full visibility, you can better plan whether to manufacture additional inventory, reassign existing inventory, or even reroute in-transit inventory to a new Destination.

5.2 Application

Consumer choice applications like retail and e-commerce have created sophisticated consumer behavior models to target advertising and promotions for specific customers and contexts. These behavior models are expected to perform better in certain functionalities like proactive purchase decisions. The maximum success of digital twin is only possible if the function of digital twin is defined clearly. As much as a consumer digital twin is concerned, complex, multi attribute behavioral models is expected to identify the future purchase behaviors of the consumers. Application of digital twin should decide the function of the digital twin. This clearly denotes the need of predictive analytics with the data collected from asset. Since the consumer choice digital twin needs decision making, the type of data collected from IoT devices should be of greater quality. Taking an identity-by-design approach builds these capabilities into the digital twin from the outset.

5.3 Enhancement

The biggest challenge while deploying digital twin for a particular asset is the evolving requirements. As the requirements grow, the data required to portray this additional requirement also grows. It becomes vital to securely scale up the data without compromising the performance. Along with this, the decision that has to be made on the consumer choice modeling is completely data driven. These data driven digital twins are generally modeled using the stochastic simulation.

5.4 System architecture

Digital twin in consumer choice modeling can become a great hit if the twin mirrors people's interests and values. If every emotion and need of a consumer is analyzed, the right product at the right time is possible. The only way to understand consumers better is through intelligent insight into their real-time data. The building of a digital twin starts with the attachment of sensors and actuators to the physical object to capture the operational data and control the object from its digital twin. The beauty of digital twin is its ability to process multiple threads of information. So along with the contextual or operational data, various historical data from CAD, ERP and browsing history are considered to build a complete picture of the object. Once the awareness about digital twin is made, and we think of implementing digital twin on a technological level, little confusion in form of the following queries may arise: What technologies do I need to make a digital twin? What are the supporting technologies already available? Do I have a model like SDLC to systematically develop a digital twin?

The typical implementation model for building digital twins can be derived from its lifecycle model and the figure below shows the phases involved in building a digital twin (Fig. 4).

5.4.1 Inception

This phase is to identify the scope of the digital twin.

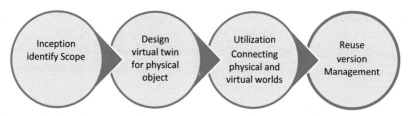

Fig. 4 System architecture.

As an example, I want to develop a digital twin for better in-store planning. The first step is to identify the scope which involves monitoring the behavior of various consumers and restructuring the store in a better way.

5.4.2 Design
Once the basic requirements are clearly understood in terms of the financial and business model, the next phase is the designing phase. For the same example, the business model is to serve the consumers better and the financial model is to reduce maintenance cost and failure reduction. A unique instance of the physical object is created in the digital world to form a digital twin. The designed model is connected with the existing data systems to derive functionalities as well as to get a full representation of the object in its environment.

5.4.3 Utilization
Once the physical object and its corresponding digital twin are defined, the next phase is the utilization phase. In this phase, getting real-time data from the physical object is the way a digital twin can help us to better model a business or finance. The communication between physical and digital twin is accomplished in this phase with the abidance from connectivity protocols and standards, security, middleware, and data storage.

5.4.4 Reuse
The digital twin is a collection of models, simulations, operational data, legacy data, and many more elements. To facilitate reuse, version management is required to maintain clear records of the digital twin.

5.5 Digital twin solutions
There are several use cases in the retail industry which can obtain surprising solutions using Digital twin. Imagine a customer enters a retail store, the

customer's need is identified automatically by a system driven by data and the system directs the customer to get everything the customer requires so that the total experience inside the stores is smooth and hassle-free. This can be made possible with a digital twin. To enable digital twin, adoption of three technologies is essential.

- Simulation tools that can replicate and virtualize the performance of physical products
- Capturing all types of data from physical assets through ultra-cheap connectivity
- A tool to provide intelligence and predictive capabilities

All these three technologies are integrated wisely by many cloud vendors like Microsoft, Amazon, and IBM. Some of the solutions that can really help one to build a digital twin for consumer choice modeling considering retail as an example would be as follows.

5.5.1 Microsoft cortana intelligence suite

Cortana Intelligence suite in conjunction with beacon technology can address numerous use cases in the retail industry. For the customers to have complete shopping experience, track the customers as they enter the store using beacon technology and the combination of this live data and database information about the customer, shops can offer targeted sales based on customer preferences.

The Cortana Intelligence Suite provides cost-effective technology services that can play an important role in enhancing the business. It is a rebranded version of the Cortona Analytics suite that can help retailers to build predictive analytics solutions utilizing the real-time data and data observed from the multiple appliances handled by the user, applications, beacon, and social media platform. Such big data will be handled, and intelligent action will be derived by Cortana.

This intelligent suite is best suited for addressing the concerns in the retail sector due to the cognitive services APIs. These are special APIs that can actually understand the method of human communication. Cognitive services not only make the system understand the person's behavior but also their needs. With such a huge potential, Cortana can successfully build solutions for retailers to:

- Push real-time recommendations
- Classifying customers
- Forecast sales
- Optimize pricing and promotions

- Enhance inventory
- Predict customer churn
- Achieve personalization with the right product at the right time

Cortana Intelligence suite operates in Azure cloud-based computing platform that can collect real-time retail transactions from millions of devices. Cortana also comes with Azure stream analytics that can process gigabytes of streaming data per second and the insights can be visualized using visualization tools like Power BI. Retailers strive hard to maintain narrow margins without compromising the customer experience. Cortana Intelligence suite services like Microsoft and Azure perform advanced analytics to generate demand forecasts to ensure the right product is in the right location at the right time.

To build an end-to-end IoT and advanced analysis solution, device management plays a crucial role. Device management includes planning, provisioning, configuring, monitoring and retirement of deployed devices. As our retail analytics includes multiple device management for streaming live data of consumers, an apt device management is the first step in the process of creating a digital twin for choice modeling. Microsoft Azure IoT does have the concept of "device twin" as a part of device management. Device twin is a JSON document that stores device metadata, configurations and conditions and it includes tags and properties.

Cortona Intelligence suite differs in the way it works compared to traditional business intelligence tools. Considering retail itself as for an example, there is a downfall of sales for a product, traditional intelligence tools try to solve the issue by understanding what and why did this happen. However, if such things must be avoided in the future what are the essential steps that should be carried out to avoid it. This type of predictive and prescriptive analytics is something essential to serve customers in a better way. This feature is available with this suite. As mentioned earlier, the input to the intelligence suite is from multiple sources and that is managed completely using the Azure device twin. Thus, the overall structure of the Cortana intelligence suite includes input, process, and output.

The type of input to the suite is the raw data collected from multiple sources like apps, databases, sensors, devices and IoT systems. These multiple sources of data are processed to get insights into the data and the output will be delivered to people, apps and automated systems. The process of converting the raw data into intelligent action happens through four stages that include multiple components to satisfy all the needs. The first stage is the information management primarily concerned with the orchestration between multiple data sources thus achieving an end to end platform. Azure data factory with its well-defined JSON scripts allows one to collect and orchestrate data from services. Another component that helps users to concentrate on insights without wasting time in search of data is the Data catalog and the ingestion activity is carried out by event hubs. The second stage components are mainly concerned with the storage of data. Cortona intelligence suite is equipped with two components that can handle and store all the types of data like structured, semi-structured and unstructured.

The third stage is the machine learning and analytics services that helps to build predictive models supporting the open source Hadoop clusters for spark, Hive, etc. as well as services for performing streaming analytics. This suite also consists of several cognitive API and SDK's to produce more personalized, intelligent and engaging insights. Power BI finally can provide powerful dashboards and graphics. Using this Cortana Intelligence suite, powerful decisions in retail can be made in sales and marketing in the form of demand forecasting, loyalty programs and customer acquisition. In finance and risk, efficient measures can be taken to detect frauds in a better way as well as good decision on pricing strategy. For customers more personalization and lifetime customer value can be made with the usage of Cortana Intelligence suite. Inventory, supply chain and store location decisions also can be made effectively by using Cortona Intelligence suite. Irrespective of numerous benefits, Cortana like suites can work effectively with integrated technologies like Tarantool that can make remarkable stand in consumer choice modeling.

5.5.2 Tarantool

Consumer choice modeling demands to store huge data as well as fast processing to get intelligent insight into the data. This fast data platform becomes possible by combining databases and application servers in memory. In memory maintenance of the database as well as application server results in the fast responses to the requests for giving the right choices to the consumers. This integration is commercially available with the name Tarantool.

Tarantool a reinstatement of multitier web applications environment. It is an integration of Lua based application server and database management system along with data grit. Tarantool is open source BSD licensed supporting Linux, Mac and FreeBSD. Tarantool can stand unique and a better choice to build a retail digital twin with its API for a database management system. Another striking feature from Tarantool is the absence of a database management system server. It works with the support of algorithms and data structures thus the storage structure is dynamic in nature and can be adopted depending on situations. With the concept of the common write-ahead log, ACID properties are achieved ensuring consistency. This ensures that it provides speed and consistency with data. Tarantool tries to give the best results for the various quality parameters like flexibility, scalability, reliability, manageability, and performance. Low latency is the greatest feature that is made possible with its in-memory storage engine. Data loss is considerably reduced with the maintenance of snapshots, even though the data is stored in RAM, it can be recovered using snapshots.

One of the beautiful aspects of Tarantool is its capability to support an agile environment and its ability to handle multiple requests simultaneously achieving greater scalability and the CPU utilization is less than 10%. Tarantool can be installed either using the docker image or through binary package which is performed with simple commands. With consumer choice modeling, the information about the consumer should be processed correctly and the digital twin should guarantee all the client requests are satisfied. System crashing is something unavoidable and that makes up a huge dissatisfaction among the consumers. This aspect gains a lot of importance and any technology that can make the services fault-tolerant and all the requests made to the API are granted. The digital twin that is built to serve the consumers might use service API's that receives services for building reports which are extremely beneficial and if that service is lost may completely disrupt the recommendations given to the consumer. The requests that aren't delivered to the service API should be handled with less downtime providing high-speed access is essential. Proxying data to Tarantool would be an ideal solution to such scenarios as it can provide a link between the processed and raw data. This shows the flexibility of Tarantool without touching the workflow and infrastructure still achieving high-speed access to data.

The processing phase of the Tarantool Data grid is a combination of multiple elements that has its rules engine, database engine, and connectivity. Tarantool holding huge cons can be a greater choice for any use case related to customer digital twin.

5.5.3 Eclipse ditto

Another well known open source technology to implement digital twin is Eclipse Ditto. Ditto possess the immense capability of mirroring innumerable digital twins residing in the digital world with the physical world.

Even though Ditto is not a full-fledged IoT platform this technology might hold some value in consumer choice modeling as this can provide support to implement software near the hardware. A typical digital twin is expected to mirror the physical asset and provide all the services around the mirrored asset. Whether the digital twin is built for industry or consumer, it should synchronize the real and digital world. All the qualities expected to build a digital twin is available in the ditto framework as it holds API for interaction, permission access and its integration with other backend infrastructure. Eclipse Ditto can be beneficial in constructing the digital twin, however, data handling and intelligence components that can serve consumer choice modeling better are still unexplored.

5.5.4 Big chain Db

Consumer choice modeling demand's security as it involves consumer's personal data as well as maintenance of various legal and regulatory compliance of suppliers. As a part of their supply chain management, retailers lease some materials from other organizations. In this scenario, it becomes essential to maintain everything about the devices that are been leased as well as the owner information. The specific agreements for the usage and other legal and regulatory compliance should be maintained. This maintenance requires a blockchain network which is not discussed in any of the above mentioned technologies. The leased devices should be monitored for the functional efficiencies using IoT and its status should be updated to Azure IoT Hub as well as legal and regulatory maintenance. This use case requires the linkage of blockchain and IoT which can be achieved with Bigchaindb.

The above technologies are just a reference to implement digital twin for consumer choice modeling. Since the consumer choice modeling focuses on the right product at the right time, various sources data both in the form of live and historical must be processed at high speeds with greater intelligence insights. Considering this perspective, Tarantool would be a great choice to create a digital twin for consumer choice modeling.

6. Conclusion

The era of satisfying a consumer through promotions, recommendations and discounts must take up a turn and scale up in such a way to understand them personally. Personalizing according to their current behavior, few days behavior and a year back should be considered before we do recommendations for them. Probably a month back, the consumer would have had interest in buying hair oils. The life of a consumer would have taken a turn in such a way that he/she completely lost hair due to a health condition and recommendation to him to buy hair oil is a great turn off and the consumer will altogether avoid visiting that shop. Such recommendations that can happen with the traditional recommendation engines should undergo a change with the abidance of the digital twin. Deeper insights on the way a customer is behaving can only be tracked with his live data. So, to build a successful behavioral digital twin, the type of data and the intelligence we get from that data holds significant. The only possible way to give the right products to the right consumers at the right time is only possible through a well-defined behavioral digital twin.

References

[1] https://medium.com/@wrld3d/digital-twin-technology-use-cases-and-how-to-build-them-11fa55b8959e.

[2] Bughin J, Chui M, Manyika J: Clouds, big data, and smart assets: ten tech-enabled business trends to watch, *McKinsey Q* 56(1):75–86, 2010.

[3] Gantz J, Reinsel D: Extracting value from chaos, *IDC Iview* 1142:1–12, 2011.

[4] Babiceanu RF, Seker R: Big data and virtualization for manufacturing cyber-physical systems: a survey of the current status and outlook, *Comput Ind* 81:128–137, 2016.

[5] https://www.monash.edu/business/marketing/marketing-dictionary/r/retail-life-cycle.

[6] Devlin B, Rogers S, Myers J: *Big Data Comes of Age (Whitepaper),* 2012, pp 1–43. Retrieved from http://www03.ibm.com/systems/hu/resources/big_data_comes_of_age.pdf.

[7] Manyika J, Chui M, Brown B, Bughin J, Dobbs R, Roxburgh C, Byers A: *Big data: The Next Frontier for Innovation, Competition, and Productivity,* 2011, pp 1–143. Retrieved from, http://www.citeulike.org/group/18242/article/9341321.

[8] https://towardsdatascience.com/disruption-in-retail-ai-machine-learning-big-data-7e9687f69b8f.

[9] Shafto M, Conroy M, Doyle R, et al. *DRAFT Modeling, Simulation, Information Technology & Processing Roadmap—Technology Area 11,* Washington, DC, 2010, National Aeronautics and Space Administration.

[10] https://www.digitalistmag.com/digital-supply-networks/2018/08/30/digital-twins-ensure-your-best-laid-supply-chain-plans-never-go-awry-06183474.

Further reading

[11] Grieves M: Digital twin: manufacturing excellence through virtual factory replication, 2014: In *White Paper,* [Online]. Available, http://innovate._t.edu/plm/documents/doc_mgr/912/1411.0_DigitalTwin_White_Paper_Dr_Grieves.pdf.

[12] Hochhalter J, et al: Coupling Damage-Sensing Particles to the Digitial Twin Concept, 2014: Accessed: Jan. 17, 2018. [Online]. Available: https://ntrs.nasa.gov/search.jsp?R=20140006408.

[13] Tao F, Zhang M: Digital twin shop-Floor: A new shop-floor paradigm towards smart manufacturing, *IEEE Access* 5:20418–20427, Sep. 2017.

[14] Glaessgen EH, Stargel DS: The digital twin paradigm for future NASA and U.S. air force vehicles. In *53rd AIAA/ASME/ASCE/AHS/ASC Structures, Structural Dynamics and Materials Conference.* 2012.

[15] Boschert S, Rosen R: Digital twin—the simulation aspect. In Hehenberger P, Bradley D, editors: *Mechatronic Futures*, Cham, 2016, Springer International Publishing, pp 59–74.

[16] Chen B, Xie Y: A computational approach for the optimal conceptual design synthesis based on the distributed resource environment, *Int J Prod Res:* 55(20):1–21, 2017.

[17] http://digitwin.ac.uk/wp-content/uploads/2018/05/Copy-of-Benchmark-April-2018-The-digital-twin-for-engineering-applications-small3.pdf.

[18] https://www2.deloitte.com/content/dam/Deloitte/cn/Documents/cip/deloitte-cn-cip-industry-4-0-digital-twin-technology-en-171215.pdf.

[19] https://dspace.mit.edu/bitstream/handle/1721.1/107989/04.Digital%20Twins.pdf?sequence=14.

[20] https://www.iosb.fraunhofer.de/servlet/is/14330/visIT_1-26-03-2018_web.pdf.

About the author

D. Sudaroli Vijayakumar is an Assistant Professor in the Department of Computer Science at PES University. She holds B.E in computer science and M.tech in Digital Communication and Networking. Her research interests span on finding machine learning based solutions in various domain ranging from software testing, wireless networks etc. She is a cisco certified professional and was a cross-functional trainer. She holds fast tracker, let's clone you awards. She is a fellow of IAENG, Analytics Society and CSI.

CHAPTER TWELVE

Digital twin: The industry use cases

Pethuru Raj[a], Chellammal Surianarayanan[b]
[a]Reliance Jio Infocomm. Ltd. (RJIL), Bangalore, India
[b]Bharathidasan University Constituent Arts & Science College, Tiruchirrapalli, India

Contents

Advances in Computers, Volume 117
ISSN 0065-2458
https://doi.org/10.1016/bs.adcom.2019.09.006

Abstract

Without an iota of doubt, the well-intended digital transformation initiatives through the smart and systematic application of digitization and digitalization technologies and tools is leading to both tactical as well as strategical accomplishments across industry verticals. Worldwide enterprising businesses ought to consistently and consciously embark on various digital innovation and disruption activities in order to be ahead of their competitors in the knowledge-driven marketplace. There are a number of pioneering digital technologies emerging to speed up and streamline the activities to meet up digital transformation goals. Digital twin is perhaps one of the widely discussed and dissected digital intelligence technologies for the forthcoming era of knowledge-filled, mission-critical and people-centric services. With the faster maturity and stability of information, communication, sensing, perception, decision-making and actuation technologies, the digital twin paradigm is acquiring a lot of mind and market shares. This chapter is specially prepared and incorporated in this book to tell the prominent and dominant use cases of the fast-evolving digital twin. We have also added a number of industry use cases and benefits for unambiguously substantiating the varied claims on the future of this unique and sustainable discipline.

1. Introduction

The emergence of software-defined cloud infrastructures and scores of integrated platforms along with a bevy of pioneering digital technologies such as machine and deep learning, streaming analytics, microservices architecture (MSA), container management solutions, the distributed and decentralized IoT architectures, fog or edge data analytics, 5G communication, etc., leads to a variety of digital disruption, innovation and transformation for the worldwide corporates and cities. The nations across the globe setting up and sustaining smarter cities are empowered with the faster maturity and stability of game-changing technologies and tools. With the continued advancements and accomplishments in the ICT (information and communication technologies) space, the speed and sagacity with which the establishment of smarter cities is really praiseworthy. The rising complexities due to the arrival and usage of heterogeneous and multiple technologies for realizing smart cities are on the climb. Therefore the adoption of complexity-mitigation and value-adding technologies helps planners, decision-makers and administrators come handy in surmounting those complications to quickly and easily bring forth people-centric, extensible, adaptive, knowledge-driven, innovation-filled, cloud-enabled, and safe cities.

Heavy industry machineries, scores of personal as well as professional devices, a bevy of humanoid robots and flying drones, a dazzling array of defense equipment, medical instruments, highly complicated machineries such as satellites, rocket launchers, earth movers, etc., the growing family of manufacturing and medical instruments, and consumer electronics are growing exponentially in widespread usage. Extra intelligence besides additional capacities and capabilities are being innately embedded into all kinds of machines and devices in order to be elegantly intelligent in their designated operations, offerings and outputs. The device ecosystem is continuously growing in order to bring in a litany of path-breaking automation for not only businesses but also for people in their everyday decisions, deals and deeds. With a number of popular miniaturized techniques, devices become slim and sleek, trendy and handy, and multi-faceted. A large number of modules are being attached with devices internally as well as externally in order to supply advanced services. There are single board computers (SBCs), which are embedded and networked.

Devices are becoming intelligent through purpose-specific and agnostic integration and orchestration. Devices are being presented as API-enabled services. Devices are being expressed and exposed as interoperable, publicly discoverable, network-accessible, and composable services. All device complexities due to the multiplicity and heterogeneity nature of devices get decimated with the service-enablement. The devices and their functionalities are hidden behind the service APIs. Device integration (seamless and spontaneous) with other devices in the vicinity and faraway cloud-based applications and data stores leads to the realization of context-aware applications.

For such devices and products, there are a couple of important goals to be fulfilled. First, it is mandatory to have a deeper and decisive glimpse of what they can do, and how, what sorts of internal as well as external risks involved in using them on day-to-day basis, how they act and react in different situations, what sort of benefits can be accrued if linked together, what is its performance level and health condition, etc., before they are actually produced. The second aspect is that manufacturers and end-users want to have the product state information continuously as the product data helps immensely in devising a workable and futuristic plan for the next version/release of the product. Multiple machines, a slew of data sources, different stakeholders, and distributed applications constantly interact with products in order to have a comprehensive and futuristic view of the products. Preventive and predictive maintenance of the products can be fulfilled through real-time data capture and crunch, which helps to extricate actionable insights in time. The remote monitoring, measurement, marketing, management and maintenance of devices can be realized.

Digital twins are the virtual replicas of physical devices [1] that data scientists and IT pros can use to run simulations before actual devices are physically built and deployed. Digital twin technology has moved beyond manufacturing and started to merge with some of the emerging technologies such as the Internet of Things, data lakes, artificial intelligence (AI) and data analytics. We primarily focus on the current and future use cases for this new technological paradigm.

2. A recap of digital twin

Digital transformation is definitely a buzzword these days. Worldwide enterprises are taking all that are needed to be succulently projected as digital enterprises. Governments across the world are seriously strategizing and precisely planning to have digital governments as a beneficial measure for their citizens. Even we hear, read and sometimes experience digital economy. Our city planners and administrators are keen to have digitally transformed cities. Thus, digitization and digitalization have led to a series of innovations and disruptions to the total human society. There are many pivotal technologies emerging for elegantly establishing digital systems and environments Digital twin is being recognized as one such technological paradigm blessed with innate strength to extensively contribute for setting up and sustaining digitally enabled organizations and institutions.

The digital twin is the virtual representation of a physical object or system across its unique lifecycle. It uses real-time data from multiple sources to intrinsically enable learning, reasoning, and contributing for extracting actionable insights. That is, digital twins facilitate two things: data-driven insights and insights-driven decisions and actions. It is all about creating a competent and assistive digital representation for all kinds of physical things, which could be anything like transport vehicles, aircraft engines, satellites and their launchers, humanoid robots, intelligent drones and turbines, complex structures such as tunnels, bridges, buildings, etc. The implementation technologies behind the formation of viable and venerable digital twins have grown substantially to include large items such as manufacturing floors, assembly lines, and even cities. There are recommendations to have digital twins for people and processes. In short, for any complicated and sophisticated system, there is an insistence for crafting an intelligent digital twin for attaining a variety of tactic as well as strategic benefits. Anyone looking at the digital twin running as a cyber application in cloud environments can cleanly glean and get crucial information to gain a deeper understanding about the corresponding physical thing at the ground level.

Concisely speaking, a digital twin is a digital representation of a physical object or system. The grandiose idea first arose at NASA: full-scale mock-ups of early space capsules and shuttles, used on the ground to mirror and diagnose problems in orbit, eventually gave way to fully digital simulations. In essence, a digital twin is a computer program that takes real-world data about a physical object or system as inputs and produces as outputs (predications or simulations of how that physical object or system will be affected by those inputs).

A digital twin begins its long life being built by specialists. Especially data scientists are to get insightfully empowered with the delectable improvisations in the digital twin paradigm. There are people who research the physics that underlie the physical object or system. They ultimately use that research output data to develop a fine-grained mathematical model that simulates the physical product in the digital space. The twin is being faithfully constructed so that it can constantly receive input from multifaceted sensors, which are externally and internally embedded in the physical object. The idea is that these specialized sensors collect behavioral, structural, external, interaction and other operational data and quickly transmit them to faraway digital twin. This allows the digital twin to perfectly simulate the physical object in real time. This offer insights into performance and potential problems. The twin could also be designed based on a prototype of its physical counterpart. This twin then can provide appropriate feedback as the product is being physically built and refined. Even a digital twin could even serve as a prototype before any physical version is constructed. Eniram is a company that creates digital twins of the massive container ships that carry much of world commerce. This is an extremely complex kind of digital twin application.

Precisely speaking, digital twin is a vital tool to help engineers understand not only how products are performing, but also how they will perform in the future. Deeper and specific analysis of the data from the connected sensors, combined with other sources of information, help us to extract vital predictions and prescriptions. With cognitive computing is all set to flourish, automated analytics for uncovering hidden patterns, bringing out useful associations, envisioning newer possibilities and opportunities, and predicting outcomes, alerts, etc., is to become persuasive and pervasive. With appropriate intelligence being squeezed out of data heaps, organizations can learn and envisage a greater and brighter future by decimating old boundaries surrounding product innovation, complex lifecycles, value creation and customer delight.

3. Digital twin key drivers

The technologies and tools for accurate sensing, perception, and vision, ambient communication, data virtualization, integrated data analytics, predictive insights, knowledge discovery and dissemination, containerized cloud infrastructures, etc., are stabilizing and maturing fast. Therefore, the concept of digital twin is growing rapidly in order to remarkably and rewardingly support and sustain digital transformation.

An inspiring example is given below. An engineer's job is to design and test products (connected cars, jet engines, healthcare instruments, manufacturing tools, etc.). The functional and non-functional requirements have to be verified and validated through automated tools, test cases and scenarios. Stress or stamina, security, scalability and availability needs ought to be fully checked. How the interactions with their environments and with owners and occupants of the environment along with remote integration also have to be thoroughly tested. An engineer testing a car component would typically run a computer simulation to understand how the component would perform in various real-world scenarios. This kind of simulation and visualization has the distinct advantage of being a lot quicker and cheaper than building multiple physical cars to test. But there are a few glitches.

The so-called computer simulations are restricted to current real-world events and environments. These can't predict how the car and its components will react to future scenarios and changing circumstances. Second, modern cars are deeply connected and increasingly autonomous. Hundreds of sensors, microcontrollers, and actuators are being attached with recent and expensive cars to offer extra convenience, comfort and choice. That is, cars are becoming software-defined. Thousands of lines of source code are being embedded in various car electronics modules. Therefore, the outdated simulation mechanism is going to fail miserably in advanced cars. This is the main reason why digital twin is garnering a lot of mind and market shares these days.

Suppose a chemical manufacturing plant produces several types of products in plenty every year. As usual, each production line has to accommodate highly specialized and expensive machines. This is supplanted with carefully detailed recipes of raw materials and machine settings to produce the end product.

Now if there is a digital copy for each of the production lines in the factory along with sensor data details collected from production line machines. This digital data is typically stored in a data historian. Then there would be the ERP data of the raw materials, production orders, and recipes. There is also quality management system data. The manufacturer has to continually optimize production yield by reducing unplanned machine downtime and minimizing the amount of by-product (widely termed as scrap) in each production run. The other important task is to lessen production quality faults, which are quite costly. If these needs are being fulfilled manually, then it is going to be a time-consuming and complex affair.

Herein comes the unique competency of any standardized digital twin software solution. This would ingest data from relevant information and operational technology (IT/OT) sources and display a virtual copy of the entire plant line through a visualization platform. Process engineers, QA teams and others can now easily understand the data in the context of the machines, the raw materials, and the entire production line environment in order to ponder about the various rationalization and optimization steps. If the digital twin solution is coupled with root cause analysis tools, then it can even point out which variables need urgent attention. This eventually speeds up root cause investigations and optimization processes.

A digital twin uses data from connected sensors to tell the current and future story of an asset. It is possible to measure specific indicators of asset health and performance, like temperature and humidity, etc. By incorporating this data into the digital twin, engineers have a 360-degree view into how the car is performing, through real-time feedback from the vehicle itself. A digital twin can help identify potential faults, troubleshoot from afar, and ultimately, improve customer satisfaction. It also helps with product differentiation, product quality, add-on services, etc.

Digital Twin Types—Forbes has articulated four distinct (but inter-related) types of a digital twin.

- *Component*—This is a digital twin of an individual component within an asset like a bulb in a scanner or a blade in a turbine. Data at the individual component level allows for data-driven decisions for operations and maintenance of that component.

- *Asset*—This digital twin is a model of an entire asset, which is a piece of manufacturing equipment, a motor or a pump. This gives a holistic view of its workings to optimize performance and enable preventive maintenance. When we have sufficient amount of data, we can avail predictive maintenance too.
- *System*—Multitude of assets form a system. A proper example is a production line in a factory. The system-level digital twin delivers data that vividly demonstrates how the dazzling array of assets team up together to achieve the intended business goal.
- *Process*—A process digital twin provides a business-level view to measure operational aspects that underpin business operations across the enterprise. It gives end-to-end visibility to optimize throughput, quality and performance of the process. It enables organizations to visualize and simulate alternative approaches to re-engineer entire processes.

Larger enterprises can ultimately land in having hundreds of digital twins. With the number of IoT devices in and around any physical twin goes up rapidly, the number of entities supplying data to different digital twins also rises considerably. This trend inspires companies to think about deploying composite digital twins that allow them to integrate all the different digital twins involved in a particular operation or process together.

In summary, digital twins enhance our ability to understand, learn, and reason from changes in physical twins and their environment. Digital twins are being touted as the stimulating foundation for producing smart applications. Digital twins continuously capture sensor data from different and distributed data sources to proactively and pre-emptively optimize performance, predict impending failures, articulate health condition, report risks, and simulate future scenarios for physical twins.

4. Digital twins for the intelligent IoT era

It is going to be the connected and cognitive era. Every commonly found and cheap thing in our everyday environments (homes, offices, hotels, hospitals, shopping malls, training halls, entertainment plazas, food joints, car parking areas, pathways, etc.) gets methodically digitized through a slew of digitization and edge technologies. Digitized items (stylishly termed as connected entities/IoT artifacts/smart objects/sentient materials) are capable of participating in mainstream computing. They are computational, communicative, collaborative, sensitive, responsive, and active. They are self, surroundings and situation-aware. Precisely speaking, we will be surrounded by a large number of digitized entities in order to contribute in realizing hitherto unheard context-aware, physical and people-centric services.

That is, all kinds of physical, mechanical, electrical and electronic systems get digitized in order to assist in producing smarter services for mankind. Digitized entities can find and bind with one another in the vicinity as well as with remotely held, cloud-hosted and cyber applications, services and databases. With these IoT devices, there arise a few crucial challenges and concerns.

1. How to remotely and rewardingly manage all these enabled devices individually as well as collectively.
2. How to produce and operate such complicated devices in a risk-free and deterministic manner.
3. How to visualize how these empowered objects collaborate, corroborate, correlate and compose to form bigger and better entities.
4. How to extricate a variety of actionable and behavioral insights out of various data being emitted by these connected devices.
5. How to do comprehensive analytics on historical and runtime data volumes being generated by one or more devices in order to extract predictive and prescriptive insights.

The concept of digital twin has started to spread its wings wider. It is being believed that this unique idea will be liberal in addressing the above-mentioned requirements with all the clarity and confidence. Let us digress into the versatile digital twin method.

As noted above, the digital twin is to give a digital representation for any physical object. Digital twin is one of the direct results of all the accomplishments in the field of the IoT paradigm. It is often said that Industry 4.0/smart manufacturing needs digital twins to really push it forward. The demand for digital twins is exploding as business enterprises and industry powerhouses grow in understanding the pragmatic applications of digital twins. Creating a digital representation of a physical object brings in a string of clear benefits. Digital twins turn out to be a strategic asset for product companies in predicting any failure, limitations, risks, etc. The productivity of industrial assets goes up remarkably with the appropriate usage of the digital twin conundrum.

It is quite beneficial to establish and use digital twins for networked and embedded systems (resource-constrained as well as intensive). These IoT devices are exponentially growing in number. That is, the multiplicity and heterogeneity of IoT assets lead to extra complexity (development, monitoring, measurement, and management). Thus, we need robust and resilient measures such as 3D modeling of those devices. Of course, modeling assets to monitor operational and maintenance performance is hardly new. The advantages with digital twin when compared with modeling are growing. There are many enterprise, cloud and mobile applications pouring in a lot of data to digital twin to do comprehensive analytics.

Further on, several sensors and actuators get attached with physical twin and this setup is bound to generate a lot of state, environmental, health condition, throughput, and interaction data, which gets transmitted to digital twin then and there. Historical and refined data are getting stocked in cloud-based data lakes. Both batch and real-time processing of all kinds of internal and external data contributes intuitively for availing advanced benefits out of the digital twin concept. The widespread acceptance of AI algorithms and approaches also copiously contribute for the greater success of this concept.

The digital twin is one of the fastest growing applications of Industrial IoT technology. It creates a complete digital replica of a physical object and uses the twin as the prime point of digital communication. Everything from manufacturing processes to sensor input, to external management software, can be fed into, and organized inside, the digital twin. Digital twins can aggregate data, making it easy to identify patterns in historical events, spot root causes, and optimize line processes.

Thus, there is a deep relationship between the digital twin (DT) discipline and the IoT idea. It is clear that the pioneering innovations and disruptions being accomplished in the IoT landscape has a greater impact on the fast-emerging and evolving digital twin phenomenon [2]. With digitized objects and assets joining in the mainstream computing, the role and responsibility of digital twins for all kinds of IoT artifacts goes up significantly. That is, not only complex machines but also simple objects are also getting their virtual representation and get benefited immensely. With the emergence of machine and deep learning algorithms along with the ready availability of powerful data analytics platforms, accelerators and engines running on cloud infrastructures, the digital twin concept gains the wherewithal to be extremely penetrative and pervasive. Industrial IoT data can be subjected to a variety of investigations to extricate actionable insights in time. Knowledge discovered and disseminated empowers the IoT device to be adaptive, autonomous and agile. Not only real-time analytics but also batch processing of IoT big data can be performed in order to visualize a lot about the device.

Digital twins can be used to predict different outcomes based on variable data. With data analytics, digital twins can often optimize an IoT deployment for maximum efficiency, as well as help designers figure out where things should go or how they operate before they are physically deployed. The more that a digital twin can duplicate the physical object, the more likely that efficiencies and other benefits can be found. For instance, in manufacturing, where the more highly instrumented devices are, the more accurately digital twins might simulate how the devices have performed over time, which could help in predicting future performance and possible failures. Thus sensor, communication, data lakes and analytics, and AI paradigms do a greater service for the betterment of the digital twin idea to flourish.

5. The levels of digital twin (DT) maturity model

There are multiple levels for digital twins. The levels typically depend on various things such as the number of sensors and actuators getting attached with physical twins and the number of different applications, services and data sources getting integrated with digital twins.

- *Partial*—At this level, the digital twin typically is connected with a limited number of data sources and sensors such as pressure, temperature and device state. This kind of twin is useful to capture a key metric or state for a low power or resource-constrained asset such as a connected light bulb that simply reports its current power condition. This level acquires enough data to create directly derived data for further analysis. For example, if pressure is down but temperature is up, and linear regression identifies a correlation, then a proper inference can be made out about the health of the asset.
- *Clone*—This form of a digital twin contains all meaningful and measurable data from multiple sources for an asset. This level is applicable when a connected asset is not power or data constrained.
- *Augmented*—The augmented digital twin enhances the data from the connected asset with derivative data, correlated data from federated sources, and/or intelligence data extracted through data analytics engine. Machine learning algorithms also play a vital role in shaping up this form of digital twin.

Thus, there are different forms and levels at which digital twins operate and deliver. With the general availability of myriad path-breaking technologies, the scope, size, and speed of digital twins are bound to vary. The sagacity and sophistication of future DTs will grow and glow. Predictive DTs will attain higher popularity with the incorporation of machine and deep learning algorithms.

6. Digital twin industry domains

As we see, there are significant impressions and improvisations through digitization and digitalization techniques, tenets and tips. A myriad business, technology and cultural transformations across a host of industry verticals ranging from healthcare, manufacturing, retail, utility, logistics, supply chain, etc., are being realized through adoption and adaptation of digital technologies. It is predicted by leading market watchers, analysts, and researchers that there will be trillions of digitized assets, billions of connected devices and millions of software services in the years to unfurl. Our everyday items are bound to be smart in their tasks and electronic systems will become smarter. People become the smartest in their decisions, deals and deeds with the unambiguous and unobtrusive assistance of multiple devices combining with one another. With digital innovations and disruptions flourishing across with the utmost nourishment by product vendors and research labs, we can safely expect and experience smarter applications and services. As accentuated before, digital twin is regarded as one of the paramount methods towards crafting intelligent systems. Many industry verticals have embraced this phenomenon in order to be ahead in their obligations to their customers, clients and consumers. This section is specially incorporated in order to tell some of the popular use cases of digital twin.

The implementation of a digital twin is an encapsulated software object that mirrors the characteristics of a unique physical object or a unique collection of physical objects. The minimum elements of a digital twin include the model of the physical object, data from the object, a unique one-to-one correspondence to the object and the ability to monitor the object. But the digital twin concept can also be applied or extended to complex entities, such as cities, enterprises and countries, to support specific financial or other decision-making processes. Furthermore, additional optional elements of digital twins, such as analytics, control and simulation, can be applied to these more abstract digital twins.

The car and the cargo vessel seem to be potential use cases for the digital twin paradigm. Objects such as aircraft engines, trains, offshore platforms, robots, drones, satellites, and turbines have to be designed and tested digitally and deeply before being physically produced and used. These digital twins come handy in doing maintenance operations efficiently. For example, technicians and operators could use a digital twin to test whether a proposed fix for a piece of equipment works well before applying the fix on the physical twin. Digital twin business applications are gaining a lot of ground these days and a number of sectors are keenly exploring the value of this concept.

- *Manufacturing* is the first and foremost area where the concept of digital twin took shape and prospered. Factories stated using digital twins to simulate their processes.
- *Automotive industry*—We increasingly read and hear about autonomous/self-driving cars. The extreme connectivity and scores of telemetry data in synchronization with the digital twin capability can help us realize next-generation connected and smart vehicles. With the continuous advancements in the DT space, the transport and logistics industries are to thrive in the days to unfurl.
- *Healthcare* sector is to benefit immensely with the digital twin idea. One prime example bandied widely is that band-aid sized sensors send health information back to a digital twin used to monitor and predict a patient's well-being.

There are other sections embracing this unique technological paradigm in order to be competitive in their offerings and operations. We will throw more light on other interesting use cases in the subsequent sections of this chapter.

7. Enterprise-scale digital twins

General Electric (GE), without any iota of doubt, is the industry pioneer in embracing and escalating the DT idea. The Digital Twin collects data from its manufacturing, maintenance, operations, and operating environments and uses this data to create a unique model of each specific asset, system, or process, while focusing on a key behaviour such as life, efficiency and flexibility). Powerful and real-time analytics is applied on the collected and cleansed data in order to uncover patterns in data volumes. The insights extracted help to detect any anomalies/outliers. Key performance metrics are maximized through analytics. The application of machine learning algorithms helps to make accurate forecasts for long-term planning. Thus, the cloud infrastructures and platforms give the required capacity and capabilities such as modeling and analytics techniques empowering any enterprise to rapidly create, tune, or modify business services for customers.

GE has built digital twins for the jet engine components that can predict the business outcomes associated with the remaining life of those components. Gas turbines can deliver the desired electrical power output at the lowest possible fuel consumption. Wind turbines collectively optimize the production of electricity from wind farms. These Digital Twins provide up-to-date and customized information that enables GE's businesses and customers to make timely decisions and intercessions for continued profitability and maximized performance.

GE is using DTs in the monitoring and diagnostics (M&D) space to flag any irregular behaviors that could be early signs of an emerging issue. Machine Learning (ML) workflows are also extensively leveraged to detect any deviation as early as possible. This finding helps to understand if that emerges as a precursor to a potential impactful event. Prediction is at the core of the DT capability and it leverages a combination of physics–based models and data–driven analytics to optimize key business indicators such as uptime and throughput. For example, the DT solution can be used to predict the remaining life of a turbine blade on a specific aircraft engine with great accuracy. This allows the application of Condition-Based Maintenance (CBM) to manage a specific engine well. It also determines the remaining life of the turbine blade after each flight or a set of flights, by evaluating operational and environmental data and customer needs.

GE has also created enterprise-scale DTs that simulate full-scale and complex systems interactions, which simulate several 'what-if' scenarios of the future and determine optimum key performance indicators for situations with highest probability. By leveraging large data sources for weather, performance, and operations, these simulations play out possible scenarios that could impact an enterprise.

There are several other enterprises focusing on establishing and using enterprise-scale digital twins in order to get benefited immensely. Gartner has popularized a new buzzword of DTO for business use cases.

Digital twins of organizations (DTO)

The concept of digital twin has percolated to several industry domains because of its innate power and serious contributions. Even enterprises and organizations are constructing enterprise-level digital twin. Enterprise architecture (EA) and technology innovation leaders have started to strategize and plan for having an appropriate digital twin for their organization. The idea is that such a digital twin eventually turns out to be a game-changer for corporate initiatives like digital transformation. The concept of a digital twin of the organization (DTO) is the adaption of the digital twin of a thing, a device or an enterprise asset. It holds the same disruptive potential to create visibility, to deliver situational awareness and to support improved enterprise decisions.

As noted earlier, digital transformation will trigger a number of business and customer-centric changes. Premium services and solutions will emerge and impact both bottom and top-line needs. Both outside-in and inside-out thinking becomes a new normal. Not only internal but also external relationships are bound to thrive. Businesses can be driven and directed in the correct path. All kinds of risks proactively identified and addressed to confidently explore new things. A DTO is a dynamic and complete software model of any organization that critically relies on operational and other decision-enabling/value-adding data to precisely and perfectly understand business vision and mission to do course-correction facilitate customer delight and heightened productivity.

7.1 Program and portfolio management (PPM)

Program and portfolio management (PPM) leaders often struggle to prioritize and monitor underlying operational scenarios of their business activities. They are also finding it difficult to adapt their program and project plans as per operational targets that keep on changing due to various reasons (business, technology, market, etc.). To fully fulfill business expectations and to set fresh goals with all the confidence and clarity, the leaders have to have the required visibility and the correct knowledge of the prevailing and evolving situation. DTO, as explained above, comes handy here in empowering leaders with all the right and relevant details in time so that they can steer their businesses in the aspired and articulated direction towards the intended destination in a stress-free manner.

A city-based organization providing the waste and recycling, water and wastewater, energy services, and biogas services has implemented a DTO to support the entire business. That is, the DTO has a strong impact on all that ranges from business strategy to operations. It has connected the operations with all running projects and programs. This arrangement brings in more productivity, customer satisfaction, employee engagement, etc.

7.2 Demand-driven value networks (DDVNs)

DDVNs are a business environment holistically designed to maximize value and optimize risk across the set of extended supply chain processes and technologies. This business environment senses and orchestrates demand based on a near-zero-latency demand signal across multiple networks of corporate stakeholders and trading partners.

A supply chain operating model is an essential part of DDVNs. Further on, there is a need for a modular and dynamic operating model that provides the much-needed a kind of operational design agility in order to respond to all kinds of changes. The operating model can help in defining standard building blocks and ways to dynamically combine them to create new outcomes.

An automotive company implemented a DTO to connect several areas in the supply network such as materials receipt, warehouse management, in-line supply, shipping and container management. This has resulted in improved overall performance through deeper visibility and enhanced collaboration over the supply network. A deeper engagement with company employees, and business partners facilitated through the DTO ultimately helped the company to achieve more with less.

7.3 Manufacturing, quality and standard operating procedures

In manufacturing, operational excellence has been insisted for decades. It is being accomplished through a slew of measures such as lean, just-in-time, productive maintenance, standardized work, total quality management, continuous improvement, etc. Such a complex requirement can be fully met only if there is a deeper visibility and situational awareness. DTO is capable of providing this.

A car manufacturer used a DTO approach to generate and maintain work instructions and job element sheets to improve safety, quality, delivery and cost. After deployment, the company experienced higher consistency of product and process quality and delivery, a more engaged workforce, more rapid problem solving, and a dramatic increase in speed of training for newcomers.

It is therefore recommended to have a DTO and its implementation framework in place for any enterprise to edge ahead of its competitors in the increasingly knowledge-driven market. Enterprising businesses are coming to the conclusion with an enhanced knowledge of DT and how it tactically and strategically impacts.

8. Digital twin (DT) industry use cases

Any new technology is being comprehensively weighed based on its broader and deeper impacts. The DT technology is gathering momentum with the praiseworthy advancements in the allied technologies such as IoT, knowledge visualization, sensing technology, artificial intelligence, data virtualization and analytics, etc. DT is being primarily used to design, test, and build next-generation intelligent systems. Any worthwhile physical entity has its corresponding virtual/logical/cyber representation, which is being meticulously built and continuously updated and managed in cloud environments. This digital companion helps to advance the functionality of physical twins and also for minutely monitoring and maintaining physical assets. The fast-evolving DT is being prescribed as a futuristic and fabulous technique for realizing the following long-term projects.

- Smart homes, hotels, and hospitals
- Smart cities, class rooms, and campuses
- Smart retail stores, warehouses, and manufacturing floors
- Smart transports buildings and smart workplaces
- Immersive augmented and virtual realities (AR and VR)

8.1 Healthcare

A digital twin can help to create a smart and safe hospital. The hospital and patient management software solutions can be thoroughly tested to understand their performance levels through the application of appropriate DTs. The quality of healthcare services being provided by hospital authorities, doctors, and nurses can be elevated sharply with the assistance of DTs. For example, a surgeon can go through a digital twin for a digital and deep visualization of the heart before embarking on the heart operation.

8.2 Smart cities

There are several subsystems for any reasonable smart city system. Drainage, and garbage management, utilities such as electricity, gas, and water, smart buildings, transport facilities, smart vehicles, entertainment centers, railway stations, airports, educational institutions, digital signage, etc., typically form a smart city system. Digital twins for these subsystems can combine into an integrated DT can serve exceptionally well for the establishment of smart cities across nations.

8.3 Digital twins in manufacturing

We are about to realize the industry 4.0 vision. For that, smart manufacturing is essential. There is a family of path-breaking technologies and tools emerging in order to simplify the path toward the vision. Digital twin is one among them. This section describes how the power of digital twin facilitates smart manufacturing. Any manufacturing environment is nowadays stuffed with an arsenal of devices, equipment, machineries, toolsets, wares, kits, pumps, etc., With the adoption of the IoT phenomenon, these entities widely deployed and used in manufacturing floors are empowered to be connected. With multi-faceted sensors and actuators being attached with these machineries, data collection and transmission get simplified. Further on, resulting products being manufactured over there are being supplemented via a number of sensors and actuators.

- *Quality management*–Minute and continuous monitoring of any product data gleaned through the attached IoT devices (sensors, tags, stickers, beacons, LEDs, chips, controllers, and actuators) has clear advantages for quality management rather than doing random inspection. Based on the collected and cleansed data, the digital twin of the product can model every aspect of the production process to proactively identify and pinpoint where quality issues may originate or occur. In addition, the digital twin contributes in pre-emptively analyzing all the aggregated data to ascertain whether there can be better materials to compose the product. The production processes also can be fine tuned to be well-optimized and organized.

- *Better product design*—All kinds of design flaws can be easily rectified through the insights generated by digital twin. Newer design can be attempted based on the actionable inputs of the digital twin. Production processes can be deeply optimized to bring in much needed efficiency. Production customization, configuration and composition can be accomplished in an agile manner with the constant contributions of the digital twin. The next version or release of the product can be a better one with a greater understanding of the users' experiences and expectations.

   Simulations come handy in visualizing fresh products and their variants. As indicated above, digital twin is capable of connecting and collecting useful data from various operational, analytical and transactional systems. Prominent enterprise, cloud, mobile and embedded applications can also be integrated with digital twin to get their data in order to do comprehensive yet real-time analytics. The digital definition of the product is enhanced with real-world performance data and this gets informed to simulation models to remarkably improve quality and integrity of designs.

- *High-performance products*—Various traits such as performance, adaptability, maneuverability, usability, etc., can be understood through digital twin and appropriate counter measures can be taken instantaneously in order to fulfill the mandated capabilities by sagaciously using the knowledge discovered by the digital twin. A kind of historical and comprehensive analytics of data of similar products throws more light on the various properties of the product and on fulfilling the key product requirements elegantly.

- *Efficient supply chain*—The digital twin can help in bringing up optimal supply chain. It enables a deeper and decisive view on materials usage and automates the replenishment process. Lean and just-in-time manufacturing can be realized through digital twin.

Many OEMs are nowadays supplying a digital twin as an incentive for each of their products in order to market well and to sell more. As a result, today we have digital twins at the individual component, system, asset and entire production process level. Large-scale and multinational manufacturing companies are not only managing a single digital twin but also, they do composite implementations of hundreds or thousands of smaller digital twins. That is, there is a management software solution needed to minutely take note of and control of ecosystems of digital twins.

- *Digital product traceability*—DT provides universal data access around a view of product systems information, or the digital thread, from requirements to design, testing, manufacturing, and visibility into the behavior of products in the field.

- *Usage-based requirements*—DT helps to analyze real-world product usage and condition data to inform feature and functionality requirements. This sharply improves fit to market and enables value-added service offerings.

- *Connected operations intelligence*—DT can aggregate and analyze, data from different and distributed assets in order to come out with unified and real-time visibility and insights toward higher performance and speedier decision-enablement with all the clarity and confidence. DT connects factory assets and ERP/MES systems to provide appropriate information toward increased operator productivity and improved production quality.

The utility, ubiquity and usability of the DT technology is steadfastly growing with different business verticals embracing this breakthrough technology. Inspiring use cases are being showcased in order to boost the application of DT. Further on, a variety of analytics, mining, and learning techniques propels this unique paradigm.

9. Digital twin applications

With a deeper understanding, the usage of this captivating discipline is spreading and rising. For example, the digital twin (DT) of an automobile prototype is a digital and 3D representation of every part of the vehicle. This is just replicating the physical vehicle so accurately that a human could virtually operate the car exactly as he or she would do in the physical world and get the same responses, digitally simulated. Not only systems but also processes are being digitally twinned. As processes are becoming complex and composed, the digital representation goes a long way in identifying any lacunae and rectifying them pre-emptively. For instance, a manufacturing plant could be modeled so that a digital equivalent can be established for each piece of equipment, operation, etc. This activity contributes for process excellence to get desired results. DTs enable smart city systems and services so that the grown city population can gain a number of composite and cognitive services in their everyday living. The city life is all set to be digitally enabled. The service versatility, robustness and resiliency are being achieved with the help of dynamic and decisive digital representations. There are other application scenarios emerging.

Physical machines are being stuffed with a variety of state-of-the-art sensors (for example, a commercial aircraft engine is being fit with 800 + sensors) for incorporating a new set of functionalities. The sensor data, when gathered and aggregated in a sensible manner, can give consolidated and accurate details on the prevailing state of the physical machine. In addition, the sensor and state data can be combined with historical data to facilitate predictive analytics to extract functional as well as non-functional aspects of the physical machine. That is, human operators are empowered with a lot of intelligence to envision and ensure the longevity and productivity of the machine.

If a remote mechanic could instantly diagnose an engine problem by consulting your car engine's digital twin and recommend that you drive to the nearest repair shop or drive your car himself remotely. This comes handy in avoiding any kind of slowdown and even breakdown. Digital twin allows collecting varied information across the physical asset's complete lifecycle and meticulously crunch it to maximize business outcomes, optimize operations and gain better return on investment (RoI). The virtual representation is the exact digital replica of any physical object.

- *Design*—The proven simulation and visualization during the design phase can be used to verify and inspect the overall 3-D design of the product and to make sure all parts fit together. The varied simulations are mechanical, thermal and electrical as well as interrelationships between these aspects.
- *System integration*—3-D visualizations on a system level can verify constraints such as spatial footprint and physical connections. By connecting to the digital twins of other components, all kinds of possible interactions can be simulated and checked. The interactions include data transfer and control functionality as well as mechanical and electrical behavior and what-if scenarios. The advantages are many. Integration effort on site and the associated downtime for the customer is sharply reduced.
- *Diagnostics*—Observation of the digital twin through a 3-D visualization can support troubleshooting. Virtual reality (VR) glasses can provide field technicians with an overlay over the real equipment to visualize parameters. Simulations can add non-observable data, such as temperatures of non-accessible parts or material stress.
- *Prediction*—Predictive maintenance has been a popular use case across industry verticals. With the faster maturity and stability of machine and deep learning algorithms, predictive and prescriptive insights are being extracted in time and used for the longevity of equipment. Whether equipment needs some rest or repair, what is the health condition and the performance level, etc. All these real-time and runtime information helps plan rational maintenance and reduce unplanned downtime.
- *Advanced services*—Digital twin in association with data analytics and AI algorithms can generate insights that can be leveraged to produce premium and breakthrough services. Devices can be self-managing, diagnosing, healing, learning, etc.
- *Adaptive field service*—It can combine real-time and historical data of any physical asset to deliver asset-specific contextual work instructions via augmented reality (AR) experiences or connected applications for increased technician efficiency.
- *Predictive monitoring service*—DT can monitor connected products minutely for any threshold anomaly. With predictive analytics capability, DT can provide real-time alerts to move from reactive to condition-based maintenance.

- *Maintenance*—A digital twin is capable of analyzing performance data collected over time and under different conditions. The knowledge thus acquired comes handy in maintaining any product intelligently. The combined analytics on both historical and runtime data supplies the required information for administrator or operator whether any component of a system needs some rest or repair. Thus, data collection and deeper investigations of data heaps ultimately empower people to take a final call on the maintenance aspect.

There are myriad opportunities to capitalize on the new capabilities of digital twins across multiple functions in the industrial enterprise. Getting started can be challenging as data initiatives around machines, processes, and environments can exist in discrete and siloes. Driving digital transformation requires having unified strategies and frameworks in place that break down siloes and combine multiple sources of data for greater efficiency and unlock higher order insights.

Digital twin is one such type of initiative or framework to help orient disparate resources toward these unified goals. But to bring these concepts to fruition, enterprises need to identify a business pain point, assemble cross-functional stakeholders, and identify a digital mission for the twin. From there, organizations need strong partners that understand challenges, such as creating the data uniformity necessary for a twin to function, or the cultural barriers to adoption. They also need technology that shortens the time to pilot and provides the necessary simulation and analytics capabilities to create value from the digital twin.

Digital twin is a centralized entity gathering data from multiple sources in order to supply right and relevant knowledge to the corresponding physical twin. The ground-breaking benefits can be realized by businesses, partners, customers, etc.

10. Digital twin benefits

All kinds of complicated and consolidated systems are bound to have their own digital replicas to gain and use a lot of complexity-mitigation tips. Not only product design and development but also product operations get simplified and streamlined to a larger extent. Digital twin vividly articulates and accentuates the various features, functionalities, and fallacies of any product, which is getting implemented. DT emerges as the centralized and converged source off actual truth adequately empowering organizations to foster beneficial collaboration across various teams and departments within any organization. It also serves ably for business partners and clients outside the organization.

Digital twin technology has the innate strength to drastically reduce the product development time and cost by up to 50% as per the reports laid down by market watchers. Through lean development cycles, increased and insightful collaboration between external and internal partners, businesses can effectively change the design, manufacturing, sales, and maintenance of complex products.

Asset-centric organizations can have an easy access to the verified and varied knowledge of assets through their digital twins. This curated knowledge elegantly contributes for real business transformation. Business strategy, planning and execution become smarter and faster with the direct contributions of the DT paradigm. The faster proliferation of IoT devices in conjunction with DT solutions promises to bring forth hitherto unheard services. The prevailing trend is that any DT-inspired and outcome-based application will get more appreciation and recognition in the days to come. With the continued growth, a dazzling array of original and situation-aware services can be composed and given to people in time.

Without any doubt, customer-centricity has been an important factor for achieving the intended success for businesses. Customer delight is the ultimate aim for any service and solution provider. DT has the competency to guarantee that for customer-facing organizations. Further on, a digital twin helps to determine the optimal set of actions that are needed to maximize some of the key performance metrics.

11. A digital twin-centric approach for driver-intention prediction and traffic congestion-avoidance

The road traffic is steadily growing due to continued growth in people and vehicle population. The connectivity infrastructure (read roads, expressways, bridges, tunnels, etc.) is not growing correspondingly and hence the research endeavors for optimal resource allocation and utilization of connectivity resources has gained a lot these days. Therefore insights-driven real-time traffic management is turning out to be an important aspect in establishing and sustaining smarter cities across the globe. IT solution and service organizations have come forth with a number of automated traffic management solutions. The primary limitation is they are unfortunately reactive and hence is an inefficient solution for the increasingly connected and dynamic city environments. Thus, unveiling real-time, adaptive, precision-centric and predictive traffic monitoring, measurement, management and enhancement solutions are being insisted as an important requirement toward sustainable cities. There are novel approaches postulated by many researchers leveraging a few potential and promising technologies and tools such as a reliable and reusable virtual model for vehicles, a machine learning model, the IoT fog or edge data analytics, a data lake for traffic and vehicle data on public cloud environments, 5G communication, etc. This section details all these in a cogent fashion and how these technological advancements come handy in avoiding the frequent traffic congestions and snarls due to various reasons.

11.1 Problem description

The various traffic statistics across cities say that the number of road accidents is on the rise, the traffic congestion is becoming alarming, the car population is growing fast, the time being spent on the roads is increasing, and the fuel and time wastage due to traffic snarl is definitely higher. On the other hand, the pleasure trip and joyride are also contributing for more vehicles on the roads. Roadside hotels and motels are increasing in numbers. The number of traffic signals is steadily growing in order to regulate the escalating traffic. There is a growing family of traffic management systems that automate several aspects.

There is a realization that for further and deeper automation, big and streaming data analytics is the viable approach and answer. There are integrated platforms (commercial-grade and open source) for enabling both the activities. These platforms are being made readily available in cloud environments. Collecting all kinds of road, car, and traffic data, carrying them to cloud platforms, subjecting the collected, curated and cleansed data to a variety of investigations in order to arrive at decision-enabling insights, taking decisions in time and plunging into appropriate actions are the major components in the workflow. However, with clouds being operated at remote locations, the idea of real-time data capture, communication, processing, decision-enablement, and actuation is out of the question. Therefore, analytics professionals are of the opinion that instead of leveraging off-premise, online and on-demand cloud infrastructures, edge device clouds are being recommended as the best fit for real-time data collection and crunching to facilitate real-time decision-making and actuation. Thus, the faster maturity and stability of IoT edge/fog computing signals the advanced traffic management capabilities.

That is, there is a high synchronization between cloud-based big data analytics and the IoT edge data analytics through edge device clouds. But then the pronounced advantages out of this design are there not to be boasted. Because the big data analytics typically does deterministic, diagnostic and historical processing and mining. That is, the processing and analytics logic have to be coded manually and deployed. Still, there are challenges in arriving at competent traffic management systems. This paper has proposed a fresh and futuristic attempt at producing viable, self-learning and automated traffic management systems.

11.2 Embarking on next-generation intelligent transport systems (ITS)

The conventional IT-enabled ITSs are found insufficient and obsolete in the increasingly connected and complicated transport world. The fast-growing traffic conundrum insists for highly sophisticated and technology-intensive solutions for the transport world. Fortunately, the technology domain is also on the fast track producing breakthrough technologies and tools for simplifying and streamlining the process towards producing highly competitive and cognitive transport systems and services. This section illustrates the famous technologies exorbitantly contributing for the faster realization of next-generation transport solutions.

Traffic lights are becoming very prominent and pervasive in urban areas for enabling pedestrians as well as vehicle drivers. There are high-fidelity video cameras in plenty along the roads, expressways, tunnels, etc. in order to activate and accelerate a variety of real-time tasks for pedestrians, traffic police, and vehicle drivers. Wireless access points like Wi-Fi, 3G, 4G, roadside units and smart traffic lights are deployed along the roads. Vehicle-to-vehicle (V2V) and vehicle-to-infrastructure (V2I) interactions enrich the application of this scenario. All kinds of connected vehicles and transport systems need actionable insights in time to derive and deliver a rich set of context-aware services. Safety is an important factor for car and road users and there is additional temporal as well as spatial services being worked out. With driverless cars under intense development and testing, insights-driven decisions and knowledge-centric actions are very vital for next-generation transports.

Every vehicle is connected. The in-vehicle infotainment system is being fit in every kind of vehicles on the road. This in-vehicle system acts as the centralized controller and gateway for the outside world. They contribute as the communication module capturing and communicating all various operational, health and performance parameter values of every significant module of the vehicle to faraway cloud environments. A cloud-hosted intelligent traffic system (ITS) has to be in place in order to act as the data cruncher, decision-maker and actuator. The ITS has to be a highly introduced.

11.3 Fog/edge analytics through device clouds

Typically cloud computing prescribes centralized, consolidated, and sometimes federated processing through a variety of cloud models ranging from public, private, hybrid, and community clouds to fulfill new-generation computing needs. Now with the accumulation of distributed and dissimilar devices emerging as the new viable source for data generation, collection, storage, and processing, the cloud idea is getting expanded substantially and skilfully toward the era of edge or fog clouds, which is a kind of distributed yet local clouds for proximate processing. That is, the growing device ecosystem of resource-constrained as well as powerful fog devices (smartphones, device and sensor gateways, microcontrollers such as Raspberry Pi, etc.) in close collaboration with the traditional clouds are emerging as the venerable force for accomplishing the strategic goal of precision-centric data analytics.

That is, the next-generation data analytics is being expected to be achieved through extended clouds, which is a hybrid version of conventional and edge clouds. That is, the sophisticated analytics happens not only at the faraway cloud servers but also at the edge devices so that the security of data is ensured, and the scarce network bandwidth gets saved immeasurably. The results of such kinds of enhanced clouds are definitely vast and varied. Primarily insights-filled applications and services will be everywhere all the time to be dynamically discoverable and deftly used for building and delivering sophisticated applications to people. There are convincing and captivating business, technical and use cases for edge clouds and analytics for discovering and disseminating real-time knowledge.

11.4 Relevant and real-time vehicle and traffic information through edge clouds

Edge analytics is gaining a lot of momentum these days. With the edge devices are being embedded with sufficient processing, storage and I/O power, they are individually as well as collectively are readied to participate in the mainstream computing. These devices can collect and process any incoming data and emit out useful information in real-time. The shared information can help the various participating sensors and actuators to plan and indulge in performing their activities with cognition, clarity, adaptivity and confidence. Vehicles on the road are being stuffed with a number of purpose-specific and agnostic sensors and actuators in order to proactively and pre-emptively capture all the right and relevant data. The centralized infotainment system or OBD dongle contributes immensely for the making of smarter vehicles. The road infrastructures are also fitted with various cameras, sensors, Wi-Fi gateways, and other electronics to enable the data gathering, aggregation and communication. The in-vehicle infotainment system readily communicates, cooperates, corroborates and correlates with the road infrastructure modules to get synched up with one another to collectively do the real-time and secure data capture, cleansing, filtering, decision-enablement and actuation.

Vehicles talk to one another as well with the roadside IT and electronics equipment to recognize and relay the real-time situation on the road. The roadside infrastructure also comprises a variety of sensors to measure the distance and the speed of approaching vehicles from every direction. The other requirements include detecting the presence of pedestrians and cyclists crossing the street or road in order to proactively issue "slow down" warnings to incoming vehicles and instantaneously modifying its own cycle to prevent collisions. Besides ensuring utmost safety and the free flow of traffic, all kinds of traffic data need to be captured and stocked in order to do specific analytics to accurately predict and prescribe the ways and means of substantially improving the traffic system. Ambulances need to get a way out through traffic-free open lanes in the midst of chaotic and cruel traffic.

11.5 Digital twin

This is the latest buzz in the IT space. The ground-level entities (physical elements) are being integrated with cloud-based applications (cyber applications). This formal integration accordingly empowers the physical entities to join in the mainstream computing. This is the overall gist of cyber-physical systems (CPSs) and the Internet of Things (IoT). Primarily, scores of industrial and manufacturing machines get integrated with remotely held applications and data sources. This setup enables the machines to be extremely an elegantly sensitive, responsive and adaptive in their actions.

Now the idea of the digital twin is to have a corresponding virtual image for a physical asset at the ground. That is, the virtual entity has all the structural as well as behavioral properties as the corresponding physical element has. The digital twin is to have a dynamic virtual/digital representation for each of the physical systems. This cloud-based virtual representation helps to gain a better and deeper understanding of all kinds of ground-level physical, mechanical, electrical, and electronics systems and how they team up to collaborate, corroborate and correlate with one another in the vicinity. The actions and reactions of these ground-level elements can be easily visualized, modeled, studied, and articulated through their corresponding virtual entities. There are other benefits of having a virtual replica of physical things. Ultimately the fresh concept of digital twin takes the current IoT capability to the next level.

11.6 The machine and deep learning methods

This is the hottest topic on the planet earth at this point in time. The data being generated and collected from different and distributed sources is growing exponentially. That is, it is the big data era. The data are simply multi-structured. The data size, speed, scope, structure, and schema are varying and the hence it is a tremendous challenge to extract useful and usable information out of big data for data engineers and management professionals. There are a number of standardized big data analytics solutions in the form of enabling tools and integrated platforms. These analytical solutions are typically performing batch processing, which is not liked by many. We are tending toward real-time analytics of big data. That is, extracting actionable intelligence in time out of big data is the motto behind the recent advancements in the analytical space. Another interesting and intriguing trend is the automated analytics. That is, next-generation analytics platforms are being stuffed with a variety of learning algorithms in order to empower the analytical platforms to self-learn, reason, train, model, understand and articulate newer evidence-based hypotheses.

11.7 Data lake for transport and traffic data heaps

Data lakes are becoming commonplace across industrial verticals. All kinds of multi-structured data gets stocked in a centralized place to be found, accessed and used for extracting useful insights out of data heaps. Data scientists are using data lakes greatly in their everyday job. For setting up and sustaining insights-driven transport management systems, data lakes are essential. We have object storage facilities in cloud environments to facilitate the realization of data lakes. Application programming interfaces (APIs) are being attached in order to open up for outside world to find and bind with data collections to envision futuristic things.

11.8 Blockchain technology

This is quite a new paradigm gaining a lot of momentum these days. This has found a lot of followers across various industry sectors. This newly introduced technology brought in newer possibilities and opportunities for the transport sector. There are forecasts that as many as 54 million autonomous vehicles will be on the road by 2035. As the number of vehicles increases so too will the volume of data. Also, by 2020 there will be 8.6 million connected features in cars and there are also estimates that there are up to 100 electronic control units in today's cars. That equates to 100 million lines of code. There are strategic use cases for the automotive industry through the fast-evolving blockchain paradigm. Smart contracts are being coded in order to bring in the required intelligence when vehicles, traffic systems and databases, drivers, owners, etc. All kinds of interactions and transactions between the various participants get securely stored through the blockchain database. Thus, in the days ahead, there will be closer and tighter integration between vehicles and the fast-growing blockchain technology.

The noteworthy factor here is that the smarter traffic system has to learn, decide and act instantaneously in order to avert any kind of accidents. That is, the real-time reaction is the crucial need and hence, the concept of edge clouds out of edge devices for collaboratively collecting different data and processing them instantaneously to spit out insights is gaining widespread and overwhelming momentum. Another point here is that data flows in streams. Thus, all kinds of discreet/simple, as well as complex events, need to be precisely and perfectly captured and combined to be subjected to a bevy of investigations to complete appropriate actions. The whole process has to be initiated at the earliest through a powerful and pioneering knowledge discovery and dissemination platform to avoid any kind of losses for people and properties. Here collecting and sending data to remote cloud servers to arrive at competent decisions are found inappropriate for real-time and low-latency applications. However, the edge data can be aggregated and transmitted to powerful cloud servers casually in batches to have historical diagnostic and deterministic analytics at the later point in time.

11.9 The proposed solution approach

We have come out with a real-time and cognitive [3] traffic congestion avoidance solution. Having studied the current lacunae in the traffic management solutions, we have come out with an advanced, extensible and

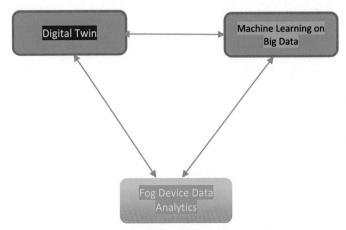

Fig. 1 Digital Twin based cognitive traffic avoidance approach.

AI-inspired solution in order to precisely and perfectly measure the traffic situation in real-time and the driver intention by leveraging the localized fog analytics, the power of the Digital twin along with the big data processing using competent machine-learning methods. The reference architecture for our solution is given in Fig. 1.

12. The solution architecture description

Following are the principal ingredients for enabling congestion discovery and dispersal, avoidance and prediction.

1. Gathering Situational Information in real time—The current road and vehicle data through fog or edge data analytics
2. Gaining Driver History, behavior and Intention through machine learning (ML) and Deep Learning (DL)
3. Data lake at Cloud for stocking historical information
4. Intelligent Transport System (ITS)
5. The Virtual Vehicle (VV) model—Digital Twin
6. Blockchain as a Service for Vehicles

The situational details are being captured through a variety of multifaceted cameras deployed along the road and route. Second, the driven intention is being captured and decided through the VV model, which was explained above in detail. The key device is the vehicle telematics system that acts as the primary gateway between the car and the outside world.

13. Edge analytics-based virtual vehicle (VV) networks

To address the traffic challenges, here is a viable proposal. With the availability of powerful cameras and sensors along the roads, bridges, expressways, tunnels, signals, etc., a massive amount of real-time as well as historical data get captured, collected, cleaned, and stocked in order to be crunched. One of the decision-enabling factors for proactively and pre-emptively avoid traffic congestion and snarl is to get the drive intention. Fig. 2 vividly illustrates how the driver intention is being deduced from the various data

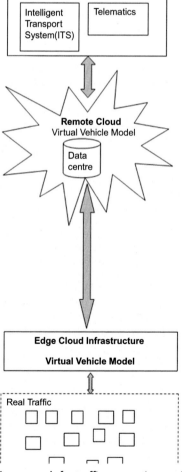

Fig. 2 Digital-twin centric approach for traffic congestion avoidance.

collection and the digital twin, which is being formed through a virtual vehicle (VV) model. The need here is to formulate a flexible and futuristic VV model in order to enable machine and deep learning algorithms to predict the driver intention with accuracy.

Since the proposed VV model makes decisions, it needs detailed driver information, such as preferences as to which lane the driver or automated vehicle is likely to select and route plans that are together considered as 'intention.' The VV model can obtain scalable, real-time driver intention data, both captured locally from the vehicle, edge cloud and from the remote cloud; by processing them in the edge and by interacting with other VVs, VVs can predict other drivers' intentions in such a way that this intention information can be used for a variety of scenarios. The VV is a virtual state of the vehicle and driver, which is processed in the edge and exists in the cloud.

The VV can interact with other VVs in the edge, where it is not limited by communication and computation resources. VVs for driverless vehicles can make decisions about path planning and interaction with other vehicles while VVs for non-autonomous vehicles can help drivers make decisions by mining other drivers' intentions. By obtaining data directly from the cloud and by actively communicating with other VVs, the VV can coordinate with others to form a VV network (VVN). The physical vehicle or traffic controller behaves like an actuator on the road, acting upon directions from the VVN to the edge.

The role of digital twin in the form of virtual representation for various physical, mechanical, electrical and electronics assets and artifacts is to grow further in the days to come. The cloud centers emerge as the best-in-class IT environment for activating and accelerating the digital twin capability in order to produce actionable insights in time. The VV model, the digital twin for the transport industry vertical, is to be realized through integration with various contributing systems in order to be adaptive.

That is, the VV model orchestrates through several entities in order to be accurate and authentic. A variety of parameters are being incorporated in order to make the VV approach viable and venerable. The localized data, being captured and filtered through edge or fog devices convey the realistic and real-time situation at the ground level. The traffic scenario, the road and the vehicle data, and other useful information are being collected by fog devices and subjected to a variety of investigations to extricate usable and useful information that can be communicated to the faraway and powerful clouds in order to synchronize with the historical data to enhance the accuracy of the decisions. The VV model comes in hand in contributing to the knowledge discovery and dissemination. Finally, the ITS acts upon based on the insights accrued.

On concluding, the transport sector is poised for accomplishing better and bigger things in the years ahead with the consistent flow of path-breaking technologies and tools. A bevy of pioneering technologies in the information, communication, sensing, perception, vision, integration, knowledge discovery and dissemination, and decision-enablement collectively are bound to do a lot of greater things for the automotive industry. There are already intelligent transport systems (ITS) and now with the addition of real-time information gathering and analytics, we can safely expect ground-breaking accomplishments for the transport and logistics industry verticals. The faster proliferation of machine and deep learning algorithms along with the evolving concept of digital twin and blockchain goes a long way in bringing more sophisticated and smarter cars, trucks, buses, trains, rocket and satellites, aeroplanes and other transport solutions. In short, it is going to be technology-splurged and software-defined world bringing immense and immeasurable benefits for every citizen of this planet earth.

14. The future

Digital twin (DT) use cases have moved out of the conceptual stage to deliver real-world impacts across enterprises, which are currently executing their digital transformation initiatives. The faster proliferation of industrial IoT (IIoT) products have laid down a stimulating foundation for the widespread interest in this captivating phenomenon. That is, multiple sensors and actuators get lavishly attached with industry machineries and instruments. Further on, other purpose-specific and agnostic devices, appliances, wares, equipment, etc., in that environment get connected with the machineries. Still on, the physical twins are integrated with a dazzling array of cyber applications and databases over any network. In short, physical assets in association with many other physical and cyber systems collect multi-structured data and transmit them to faraway digital twin to squeeze out actionable insights and context-awareness information in time.

Coupled with increasingly data aggregation, simulation and knowledge visualization capabilities, new insights can be revealed to improve operational effectiveness, differentiate products and services, and increase worker productivity. And with augmented reality emerging as the new HMI to bring 3D content and real-time insights to workers, the projected era of digital twin is to see the reality.

Digital twin (DT) is already being leveraged by various industry houses to stay ahead of digital disruption by understanding changing customer preferences, market sentiments, technology advancements, etc. This technological paradigm is capable of bringing in radical changes to businesses such as delivering quality products for the knowledge-driven market quickly and extracting viable intelligence out of product data through DT to produce better and customer-aligned products. Thus, product intelligence is the crucial difference primarily achieved through the distinct competencies of the digital twin paradigm. The future is also bright.

The powerful emergence of artificial intelligence (AI) and cognitive computing can substantially increase the special abilities of the digital twin. Technologies such as machine and deep learning algorithms, computer vision, natural language processing (NLP), real-time analytics of big data, etc., are recognized as the futuristic thing for next-generation digital twins. Automated analytics is the new paradigm with the pioneering advancements in AI and with the combination of AI and digital twin is going to be game-changing for product engineering and advancement.

15. Conclusion

A digital twin is a dynamic virtual representation of any digitized entity. Mechanical and electrical systems are being digitized with the systematic application of edge and IoT technologies. Consumer, automotive and industry electronics, avionics, robotics, and other electronic systems are the natural products to have their own digital twin.

Digital twin is typically a software package or library getting hosted and managed in cloud infrastructures. Digital twin is continuously empowered as it gets current occupational and operational data of the physical twin. Historical and curated data from multiple data sources such as databases, data warehouses and lakes, etc., can be pumped to digital twin to arrive at actionable insights. The other motivations for the overwhelming acceptance and adoption of the digital twin concept include the unprecedented success of data analytics discipline. We are being bombarded with a number of highly matured and stabilized data analytics engines, accelerators and platforms. In the recent past, machine and deep learning algorithms are being widely used for extricating predictive and prescriptive insights in time. Thus, all kinds of decision-enabling product data, technologies such as artificial intelligence (AI) algorithms and data analytics platforms, architectural patterns such as microservices architecture (MSA), infrastructure optimization through containerized cloud environments, etc.

The capability to refer to data stored in different places from one common digital twin directory enables simulation, diagnostics, prediction and other advanced benefits.

References

[1] https://www.altoros.com/blog/optimizing-the-industrial-internet-of-things-with-digital-twins/.
[2] https://arxiv.org/ftp/arxiv/papers/1610/1610.06467.pdf.
[3] https://cognitiveworld.com/articles/emergence-cognitive-digital-physical-twins-cdpt-21st-century-icons-and-beacons.

About the authors

Pethuru Raj working as the Chief Architect in the Site Reliability Engineering (SRE) division, Reliance Jio Infocomm Ltd. (RJIL), Bangalore. The previous stints are in IBM Cloud Center of Excellence (CoE), Wipro Consulting Services (WCS), and Robert Bosch Corporate Research (CR). In total, I have gained more than 18 years of IT industry experience and 8 years of research experience. Finished the CSIR-sponsored PhD at Anna University, Chennai and continued with the UGC-sponsored postdoctoral research in the Department of Computer Science and Automation, Indian Institute of Science, Bangalore. Thereafter, I was granted a couple of international research fellowships (JSPS and JST) to work as a Research Scientist for 3.5 years in two leading Japanese universities. Published more than 30 research papers in peer-reviewed journals such as IEEE, ACM, Springer-Verlag, Inderscience, etc. Have authored and edited 20 books thus far and focus on some of the emerging technologies such as IoT, Cognitive Analytics, Blockchain, Digital Twin, Docker-enabled Containerization, Data Science, Microservices Architecture, fog/edge computing, Artificial intelligence (AI), etc. Have contributed 35 book chapters thus far for various technology books edited by highly acclaimed and accomplished professors and professionals.

 **Chellammal Surianarayanan** is an Assistant Professor of Computer Science at Bharathidasan University Constituent Arts and Science College, Tiruchirappalli, Tamil Nadu, India. She earned doctorate in Computer Science by developing computationally optimized techniques for discovery and selection of semantic services. She has published research papers in *Springer Service-Oriented Computing and Applications, IEEE Transactions on Services Computing, International Journal of Computational Science, Inderscience,* and the *SCIT Journal of the Symbiosis Centre for Information Technology,* etc. She has produced book chapters with IGI Global and CRC Press. Recently she produced books on *Cloud computing with Springer* and on *MicroServices Architecture with CRC Press.* She has been a life member of several professional bodies such as the Computer Society of India, IAENG, etc. Before coming to academic service, Chellammal Surianarayanan served as Scientific Officer in the Indira Gandhi Centre for Atomic Research, Department of Atomic Energy, Government of India, Kalpakkam, Tamil Nadu, India. She was involved in the research and development of various embedded systems and software applications. Her remarkable contributions include the development of an embedded system for lead shield integrity assessment, portable automatic air sampling equipment, the embedded system of detection of lymphatic filariasis in its early stage, the and development of data logging software applications for atmospheric dispersion studies. In all she has more than 20 years of academic and industrial experience

Machine learning and deep learning algorithms on the Industrial Internet of Things (IIoT)

P. Ambika
Kristu Jayanti College, Bangalore, India

Contents

Advances in Computers, Volume 117
ISSN 0065-2458
https://doi.org/10.1016/bs.adcom.2019.10.007

Abstract

Deep transformation and human progress is a new industrial revolution that makes "Automation of Everything." It connects all digital interfaces, data analysis and control of the physical world through networks of computers. This key revolution promises everyone to unlock trillions of opportunities in the next decade. Human could feel massive improvements in productivity in physical and digital industries that enhances quality life of a human healthier and more sustainable community. In the world of IIoT, the creation of massive amounts of data from a various sensors is common and there is lot of challenges. This goal of this chapter is to provide a comprehensive review about Machine learning and deep learning techniques, popular algorithms, and their impact on Industrial Internet of Things. This chapter also delves use cases where machine learning is used and to gain insights from IoT data.

1. Introduction

Industry 4.0 revolution forcing every industry and its applications require change that rely heavily on operational technology (OT), such as manufacturing, transportation, energy, and healthcare. Previously, Industrial IoT needs fog and edge computing [1] technologies to provide the necessary connectivity within Industry 4.0. But this revolution brings another interconnected component that is essential for IIoT: Analytics.

Machine learning and deep learning algorithms enhance the ability of big data analytics and IoT platforms add value to each of these segments. Industrial IoT deals with three different types of IoT data.

1. Raw (untouched and unstructured) data.
2. Meta (data about data).
3. Transformed (valued-added data).

These algorithms will support that manage each of these data types in terms of identifying, categorizing, and decision-making. Deep Neural Networks coupled with big data analytics that draw meaningful and useful as information for decision-making purposes. The use of Artificial Neural Network in IoT and data analytics will be crucial for efficient and effective decision-making, especially in the area of streaming data and real-time analytics associated with edge computing networks.

Industrial Internet of Things (IIoT) applications include many industry verticals like healthcare, retail, automotive, and transport. IIoT will significantly improve reliability, production, and customer satisfaction in many industries. IIoT will initially improve existing processes and augmented current infrastructure but the ultimate goal is realize entirely new and intensely improved products and services.

Many industries understand how and where IoT technologies and solutions will drive opportunities for operational improvements, new and enhanced products and services, as well as completely new business models. Machine learning and deep learning algorithms on IIoT will significantly improve reliability, production, and customer satisfaction that will rely upon integrating key technologies, devices, software, and applications. Everything needs a substantial breadth and depth of technologies that require careful integration and orchestration.

These technology makes machine smarter that represent intelligent devices, machinery, equipment, and embedded automation systems [2,3] that perform monotonous tasks and can also solve complex problem without human intervention. It should also ensure smartness in enterprise include improvements in the smart workplace, smart data discovery, cognitive automation and many more. A digital twin is intended to be a replica of any physical assets or processes, etc. Mostly it is referred as an outcome of internet of things (IoT) that exponentially expanding all these devices and provide us with an equally expanding data and can also be analyzed for its efficiency, design, maintenance and many other factors. An important and significant factor about any digital twin is to continuously update and "learns" in near real-time any change that may occur. We are witnessing IoT paradigm and its solutions made lot of changes in the industry market place.

This chapter evaluates various machine learning and deep learning algorithms and their use relative to analytics solutions that rapidly grow in enterprise and on industrial data arena. It also assesses emerging business models, its solutions and provides forecasting for unit growth and revenue for both analytics and IIoT. In particular this chapter focuses on the challenges associated with the energy efficiency, real-time performance, interoperability, and security. It also provides a systematic overview of the state-of-the-art research efforts and potential research directions to solve Industrial IoT challenges.

2. IIoT analytics overview

Industrial IoT architecture involves four key components

Things—Actual machines or systems that are involved in the industry to be monitored by collecting their data. It is the real source of data.

Intelligent Edge Gateway—It is a software component that closely associated with the edge devices that is capable of collecting, aggregating, sanitizing streaming of light data. It allows us to push aggregated and relevant result to the IoT cloud. Generally, it acts as a mediator between the things and the cloud IoT platform.

IoT Cloud—Core IoT platform that capable of handling enormous amounts of data with data analytics, machine learning, and artificial intelligence techniques. Device management, stream analytics, events processing, rules engine, alerts and notifications are the processing capabilities. It includes big data, machine learning capabilities and also other services such as authentication, multitenancy, end-to-end security, SDKs, and platform APIs.

Business Integration and Applications—It is a backend system that integrates several IT systems to ensure machine data to complete the circle of operations. Examples of such systems are ERP, QMS, Planning and Scheduling, etc. (Fig. 1).

Data analysis can be divided in to three types depends on the type of result it produced. They are descriptive analysis, predictive analysis and prescriptive analysis (Fig. 2).

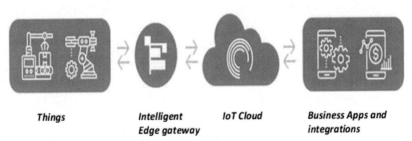

| Things | Intelligent Edge gateway | IoT Cloud | Business Apps and integrations |

Fig. 1 Ideal IIoT analytics architecture.

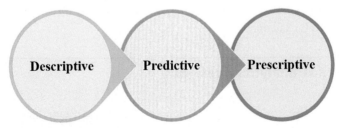

Fig. 2 Types of data analysis.

Descriptive data analysis defines what is happened/happening; we can pull data to gain insight in to current operating condition. It involves summarizing, organizing the data, describe the data but do not attempt to make inferences from the sample.

Predictive analysis foresees problems before they occur. It is a branch of advanced analytics that is used to make predictions about unknown future events.it uses techniques from data mining, statistics, modeling, machine learning, and AI to analyze current data to make predictions about future.

Prescriptive analysis is one step ahead of predictive analysis and recommends solutions for upcoming problems. It suggests decision options or mitigates a future risk, and also illustrates the implications of each decision option. It automatically process new data to improve the accuracy of predictions and provide better decision options.

3. Types of data

Machine data does not reveal complete insight in every case. By combining data from different sources we can create new insights. It is important to define a few key concepts related to data and there are two categorization of data in IoT perspective whether the data is structured and unstructured.

Structured data has a schema or model that defines how the data is organized in a simple tabular form. This type of data found in computing systems (Fig. 3).

Structured data resides in relational database and data may be human or machine generated as the data is created within the RDBMS structure. Human generated queries used to extract useful information. Some of the popular applications using structured data like railway reservation systems, sales transactions and ATM activity.

3.1 Unstructured data

Unstructured data does not follow logical schema that decode the data through traditional programming. This data does not fit in to a predefined data model. The data may be textual or non-textual and human generated. Unstructured data includes text files, email, social media, website, mobile data, media, satellite images, scientific data, sensor data, etc.

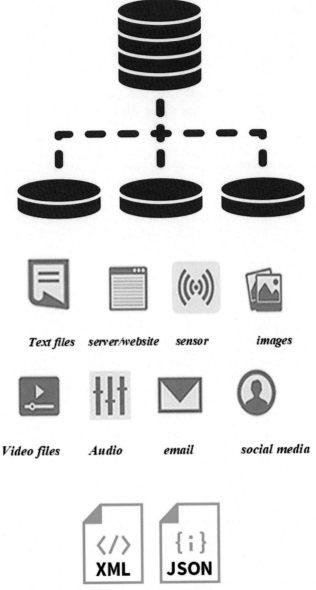

Fig. 3 Structured, unstructured and semi-structured data.

3.2 Semi-structured data

Semi-structured data is a combination of structured and unstructured data and shares characteristics of both. It also follows certain schema, consistency and exist to ease space, clarity. CSV, XML and JSON documents are semi-structured documents. NoSQL databases are considered as popular to handle semi-structured data.

IIoT devices generate structured, unstructured and semi-structured data. Structured data can be easily managed and processed because of sell defined structure. On the other hand, semi-structured and unstructured data require data analytics tools for preprocessing and managing. Most of the IIoT networks data is in transit or motion, for example, web browsing file transfers and email. In industrial IoT perspective data from edge device passes in the network always in motion. It may be filtered and processed by some other devices connected in the same network or it may sent to data center. When it comes to data center it could be processed by the real-time data analysis software and the responses sent back to devices.

4. Challenges in IIoT

4.1 Security

Security is a challenge in IoT because of two characteristics: one is the devices in IoT are constrained and the second is the scale. A device is constrained when it has limited processing power, memory, battery and bandwidth. The constrained devices cannot protect themselves given their limited computing and battery capabilities. The devices are also constrained in terms of standard interfaces they expose for managing them. The constrained devices are easy target for DDoS attacks and need an external support.

4.2 Interoperability

To address the device diversity on IoT, connected devices should follow same language of protocols and encodings. This could be addressed by an interoperable IoT environment. Practically it is complex in systems that has different layers with communication protocols stack between devices. Interoperability may also increase the economic value of the IoT market. Purchase of IoT products and services also depends on interoperability. Interoperability will be a significant consideration that enable connected devices to work in a better environment.

4.3 Real-time response

Some applications require real-time response, e.g., ability control temperature of boiler. And, IoT infrastructure is still evolving, and is even more challenging in developing countries to be able to maintain connectivity to the cloud all the time. An IoT application should continue to function even if the connectivity to the cloud breaks.

4.4 Future readiness

IoT is still evolving and requires the current deployments should support future application and new application service providers. The success of IoT deployments depends on the ability to allow innovation over the existing infrastructure.

To overcome the above challenges and to enhance edge computing capability, context awareness is the key component that needs to be enabled with the use of machine learning algorithms.

5. Need of contextual analysis in IIoT

Authors said [4–6] IIoT is the result of convergence between Operational Technology (OT) and traditional information and communication technologies (ICT). Traditional approach lacks in understanding of the differences between conventional enterprise ICT systems and the cyber-physical systems that are found in industrial applications. In contrast some assessment [7] of the security challenges faced in securing IoT devices recognized some of the constraints. It indicates current security mechanisms, based on "traditional public-key infrastructures will almost certainly not scale to accommodate the IoT's amalgam of contexts and devices."

The researchers were taking into account the limitations of IIoT devices, for example, sensors, actuators and RFID tags, where there are processing, power and economic constraints limiting the use of strong encryption. Adding to that, there is also the mismatch between the relative short lifespan of many devices and the relative longevity other devices, which are often expected to have a lifespan an order of magnitude longer than their counterparts.

The previous section focused on issues with IoT devices, such issues prevents any device from communicating, or simply that the device's battery has drained and is in need of replacement/replenishment. In real-time responsive analysis of any industrial device expected without any issues, but there's certain information that's important to highlight for everyone.

For example, in agriculture devices attached to mobile agricultural equipment (e.g., a tractor) that enable that equipment to be tracked. The devices will likely be using GPS to get location data for the tractor, but unfortunately, GPS doesn't work effectively when vehicles are inside buildings.

If all you show on the UI is the last known location of the tractor, and that tractor is now being stored inside a building, the GPS location won't reflect the actual location of the tractor and can confuse users and geofences too.

In this scenario, there are no issues with the device itself. However, the contextual state that the device is in (i.e., inside a building) is important to the functionality of the IoT solution. Contextual data is the key component to classify the device—and by extension, the tractor—into a state (e.g., "indoors"). It always provides helpful information to everyone so they're not confused when the device can't get an accurate GPS position inside a building. If the device itself can't tell when it's inside. All the device knows is that it isn't getting a good signal from GPS satellites and therefore can't acquire an accurate GPS position. This section explore some of the key aspects of contextual IoT device management using machine learning and to manage devices contextually if you're building, buying, and/or implementing massive-scale IoT solutions.

6. Machine learning for contextual analysis

Machine learning plays a major role in the contextual analysis in IIoT that can help the system getting better in analyzing and troubleshooting the network issues and issues with IoT devices, over time.

Some of the capabilities of machine learning which can be developed are:
- Prediction models
- Anomaly detection
- Recommendation models

Root Cause Analysis (RCA) capabilities aims to find the root cause of any issues by a great accuracy by analyzing the logs and historical events. It is very important to find out the root cause in an IoT environment so that no critical system is out of service for long time. Manual identification of the root cause can take hours or sometimes even days. Machine learning based Root Cause identification system can do it in minutes or sometimes in seconds. The system works on live data as well as on the historical data and learns over time to get better in recognizing patterns and correlating them back to the root cause. This is a very powerful capability which can go a long way in manageability and usability of IoT deployments.

"Rule-based Analysis" can be the starting point of getting into the machine learning territory. The most important thing is to develop a Rules Engine, which can be decoupled from the implementation deep down.

This decoupling can help even the most novice users of the IoT systems to enhance the system by adding new rules.

Security models based on these concepts can help find out and flag hidden and future threats. Once the system has historical data and trends, the context feeds to itself resulting in self-learning. It is very important to train such system for contextual analysis over a large number of network and deployment scenarios. This would help the system giving much accurate predictions and recommendations.

7. Role of analytics in IIoT

An Analytics System is very important for working with a very large amount of data (big data) and creates meaningful reports and heat maps to take decisions. The following list covers high level user visible requirements for an analytics system:

1. Data correlation to find the overall picture
2. NBI feed in JSON format for adopters to consume the data—resulting in better adoption and reporting. SCAP (Security Content Automation Protocol) based reporting can also be used
3. Data funnels, filters and segmentations—important to slice and dice the data we have
4. Both structured and unstructured data analytics—this is very important as insights into security violation can come in any form.

All of these along with machine learning capabilities help in the contextual analysis of the system resulting in better security.

7.1 Machine learning algorithms

Machine learning algorithms can be applied on IIoT to reap the rewards of cost savings, improved time, and performance. In the recent era we all have experienced the benefits of machine learning techniques from streaming movie services that recommend titles to watch based on viewing habits to monitor fraudulent activity based on spending pattern of the customers. It can handle large and complex data to draw interesting patterns or trends in them such as anomalies. Machines are needed to process information fast and make decisions when it reaches the threshold. There are many machine learning algorithms listed in Table 1 that help to do better data analysis in industrial IOT devices.

Table 1 Machine learning algorithms.

Algorithm	Type of the task
K-nearest neighbor	Classification
Naïve Bayes	Classification
Support vector machine	Classification
Linear regression	Classification/Regression
Random forest	Classification/Regression
K-means	Clustering
Principal component analysis	Feature extraction and dimensionality reduction
Canonical correlation analysis	Feature extraction
Neural networks	Classification/Regression

7.2 K-nearest neighbor algorithm

K-nearest neighbors (KNN) [8] is to classify a new given unseen data point by looking at the K given data points in the training set that are closest to it in the input or feature space. We use distance metric, such as Euclidean distance to find the K nearest neighbors of the new data point. To formulate the problem, let us denote the new input vector (data point) by x, its K nearest neighbors by $N_k(x)$, the predicted class label for x by y, and the class variable by a discrete random variable t.

8. Naïve Bayes algorithm

Given a new, unseen data point (input vector) $z = (z_1, z_2, ..., z_m)$, naive Bayes classifiers, which are a family of probabilistic classifiers, classify z based on applying Bayes' theorem with the "naive" assumption of independence between the features (attributes) of z given the class variable t.

Naive Bayes classifiers require a small number of data points to be trained, and can deal with high-dimensional data points, and also are fast and highly scalable [9]. Popular applications that use this algorithm are spam filtering [10], text categorization, and automatic medical diagnosis [11].

9. Support vector machine (SVM)

It is probabilistic based technique that does binary classification and aim to find the dividing hyper plane that separates both classes of the training set with the maximum margin. The predicted label of a new, unseen data point is determined based on applying machine learning algorithms to Internet of Things use cases. Some of the application that use support vector machine (SVM) are Anomaly Detection, Smart Traffic, Smart Environment Traffic Prediction, Finding Anomalies in Power Dataset [12–14].

10. Linear regression

Multiple Linear Regression is a very simple approach for predicting a quantitative response Y on the basis of multiple predictor variables X_p. In general, suppose that we have p distinct predictors. Then the multiple linear regression models take the form

$$y = \beta_0 + \beta_1 x_1 + \beta_2 x_2 + \varepsilon$$

11. Random forest (RF)

Random forest [15] is an ensemble learning method used for classification similar like that build by constructing multiple of decision trees at training time and outputting the class that is the mode of the classes (classification) or mean prediction (regression) of the individual trees.

The Random Forest model is also efficient on larger data sets, and can handle thousands of input variables.

12. K-means clustering

K-means algorithm [16] is to cluster the unlabeled data set into K clusters (groups), where data points belonging to the same cluster must have some similarities. In the classical K-means algorithm, the distance between data points is the measure of similarity. Therefore, K-means seeks to find a set of K cluster centers, denoted as $\{s_1, \ldots, s_k\}$, such that the distances between data points and their nearest center are minimized. It has many limitations because of the use of Euclidean distance as the measure of similarity and also types of data variables that can be considered. Moreover cluster centers are not robust against outliers. Authors [17] have used K-means in real-time event processing and clustering algorithm for analyzing sensor data that uses open IoT middleware as an interface for innovative analytical IoT service.

13. Principal component analysis (PCA)

Principle component analysis (PCA) technique is used for feature reduction. Actually, what PCA does is feature extraction in health analytics, if we know that certain features are very strong in predicting Coronary Artery Disease patients, the doctors need to know what these features are. In order to reduce the number of feature to make the feature recommendation task tractable PCA [18] is used and it does not alter the original features.

Data is grouped in different clusters, which are usually chosen to be far enough apart from each other spatially, in Euclidean distance, to be able to produce effective data mining results. Each cluster has a center, called the centroid, and a data point is clustered into a certain cluster based on how close the features are to the centroid-means algorithm iteratively minimizes the distances between every data point and its centroid in order to find the most optimal solution for all the data points.

14. Canonical correlation analysis (CCA)

It is a linear dimensionality reduction technique related to PCA that deals with one variable but CCA deals with two or more variables. Its objective is to find a corresponding pair of highly cross-correlated linear subspaces so that within one of the subspaces there is a correlation between each component and a single component from the other subspace. The optimal solution can be obtained by solving a generalized eigenvector problem [19]. Authors have [20] used PCA with CCA for detecting intermittent faults and masking failures of indoor environments.

15. Neural networks

Multi-dimensional requires a lot of computing power and it is difficult to determine the set of parameters to input. Neural networks mimic the way the human brain works, just like how human recognize scenes from the environment that recollects the distinct shapes. There are different types of neural networks and is used in many architectures, applications and use cases.

Deep Neural network types are:
- Convolutional Neural Network
- Recurrent Neural Network
- Deep Belief Network

Deep learning is extended effective approach on machine learning, and exhibits its strength when a system "automatically" extracts features such as specified colors and shapes from a massive amount of data consisting of voice, images and other elements. Features have been extracted manually based on a trial-and-error with the help of the knowledge provided by experts in respective fields in most of the machine learning tasks and deep learning allows extraction, classification and inference entirely automatically through the learning features themselves with high accuracy. AI technology has grown phenomenally by deep learning.

Many industries have undertaken the development of its AI technology for many years and applying this technology into services for industrial IoT. For example, communication AI system supports human activities by recognizing human's intentions and situations through complex information such as voice and images. Expanding the use of deep learning for the industrial IoT region to improve the "speed" and "accuracy" in data analysis utilizing experiences and knowledge cultivated with many industrial applications. In addition to voice and image data, IoT data corrected by many kinds of sensors which are installed in industrial equipment and products are also analyzed through machine and deep learning system which is based on learning, inference and action. They decide to improve the quality of customer products and services, as well as the business operations of its customers.

Deep learning is also greatly contributing to digital twin creation. For example, monitoring and inspecting system for electric power infrastructure using a drone system collaborated with Alpine Electronics, Inc. and Toshiba. Alpine Electronics, Inc. works based on image recognition technology. The image recognition technology that uses deep learning is adopted to find the damage on the power transmission lines. The new system allows detection of places with a damaged part quickly with high accuracy by examining images photographed by the drone.

15.1 Dynamic rules using machine learning and deep learning

The context is the real world scenario that arises based on the present environment of the user. A rule is defined as a set of conditions for that context. The real-world scenario is translated into a context that is defined by the rules that mentions the appropriate action that is to be taken. Based on the rules, the user takes the required decisions and triggers the action real-time. The decision can, however, be automatic too. Once a model

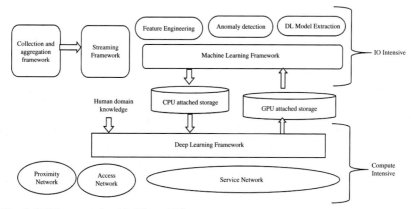

Fig. 4 IIoT framework with ML and DL.

exists (or is created over a period of time), and the context complies with the model, a rule-based decision is quite simple to take.

One of the ways this can be done is, to create a model over time and then define a "baseline." A baseline, from then onwards, can be used to compare every data points (Fig. 4).

Few of the machine learning related requirements for such a system, where it can be very relevant, would be as follows:

1. The system should be able to do Data Correlation using various rules.
2. The system should be able to work with different depths of analytics based on the requirement.
3. The system should support Anomaly Detection which would come handy when the data point does not comply with the model. In conventional model, the first reaction of the system would be to discard such data. However, a machine learning system should at least try to find the cause of such anomaly. This can give a lot of insights into the context where these capabilities can take corrective actions.
4. Such system should also possess the capability to perform Predictive Analytics. This capability would keep the system ready to tackle the upcoming situations, which can mean a threat or a condition which requires attention.
5. The system should be able to show the Trend over a period of time.

15.2 Domain intelligence using ML and DL

Operations of ML and DL classified in to two groups, one is at the edge—data are collected and processed locally either on the device or on the gateway. Second is at a central computing unit such as cloud. Depends *on where the IIoT data is processed we can derive intelligence* in to four domains.

16. Monitoring

Data on a smart object is processed based on conditions and it refers to external factors such as temperature, humidity or presence of dioxide, etc., ML and DL techniques used to monitor that allows us to early detect failure conditions.

17. Behavioral control

It always work with monitoring. A threshold value is defined along with the given set of parameters. Machines dynamically learns and alerts if there are deviations.

18. Optimization

Deep neural networks can be combined with the above such multiple systems to estimate some chemical mixture for target air temperature. Objective of such layered architecture is to optimize the result.

19. Self-healing

Popular approach in deep learning is closed loop. Systems that are capable of doing operation optimization can also be programmable to dynamically monitor and combine new parameters that deduce and implement new optimizations. Then the system become self-learning or healing.

20. Summary

Data is growing exponentially with the arrival of the Industrial Internet of Things and to acquire more data, more advanced technologies are required to scrutinize and filter out the important information and value held within. Machine learning algorithms and deep learning techniques allows us to extract large amounts of data, identify patterns and trends within it, and make predictions. Industrial IoT decision-making would be a greater advantage to optimally control and manage their assets example digital twins. Preventing such assets from failing, forecasting demand, and optimizing movements of the workforce are prevalent that are well managed by machine learning techniques and Deep Neural Networks and also improve the overall performance of the organization. By combining these elements and more on Digital Twins would result in significant benefits.

References

[1] Aazam M, Huh E-N: Fog computing micro datacenter based dynamic resource estimation and pricing model for IoT. In *IEEE 29th International Conference on Advanced Information Networking and Applications*, 2015, IEEE, pp 687–694.

[2] Atzori L, Iera A, Morabito G: The internet of things: a survey, *Comput Netw* 54(15):2787–2805, 2010.

[3] Cecchinel C, Jimenez M, Mosser S, Riveill M: Architecture to support the collection of big data in the internet of things. In *IEEE World Congress on Services*, 2014, IEEE, pp 442–449.

[4] Steenstrup K, et al: *Predicts 2017: IT and OT Convergence Will Create New Challenges and Opportunities,* 2016, Gartner.

[5] Hertzog C: *Smart Grid Trends to Watch: ICT Innovations and New Entrants,* 2012, OECD Publishing. https://doi.org/10.1787/5k9h2q8v9bln-en.

[6] Shah N: *IT and OT Convergence—The Inevitable Evolution of Industry,* 2017, IoT for All.

[7] Roman R, Najera P, Lopez J: Securing the internet of things, *IEEE Comput* 44:51–58, 2011. https://doi.org/10.1109/MC.2011.291.

[8] Cover T, Hart P: Nearest neighbor pattern classification, *IEEE Trans Inf Theory* 13(1):21–27, 1967.

[9] McCallum A, Nigam K, et al: A comparison of event models for naive bayes text classification. In *AAAI-98 Workshop on Learning for Text Categorization*, Vol. 752, 1998, Citeseer, pp 41–48.

[10] Metsis V, Androutsopoulos I, Paliouras G: *Spam Filtering with Naive Bayeswhich Naive Bayes?* 2006, CEAS, pp 27–28.

[11] Webb GI, Boughton JR, Wang Z: Not so naive bayes: aggregating onedependence estimators, *Mach Learn* 58(1):5–24, 2005.

[12] Ni P, Zhang C, Ji Y: A hybrid method for short-term sensor data forecasting in internet of things. In *11th International Conference on Fuzzy Systems and Knowledge Discovery (FSKD)*, 2014.

[13] Shukla M, Kosta Y, Chauhan P: Analysis and evaluation of outlier detection algorithms in data streams. In *International Conference on Computer, Communication and Control (IC4)*, 2015, IEEE, pp 1–8.

[14] Shilton A, Rajasegarar S, Leckie C, Palaniswami M: Dp1svm: a dynamic planar one-class support vector machine for internet of things environment. In *International Conference on Recent Advances in Internet of Things (RIoT)*, 2015, IEEE, pp 1–6.

[15] James G, Witten D, Hastie T, Tibshirani R: *An Introduction to Statistical Learning,* 2013, Springer.

[16] Coates A, Ng AY: Learning feature representations with K-Means. In Montavon G, Orr GB, Müller KR, editors: *Neural Networks: Tricks of the Trade. Lecture Notes in Computer Science*, Vol. 7700, Berlin, Heidelberg, 2012, Springer.

[17] Hromic H, Le Phuoc D, Serrano M, et al.: Real time analysis of sensor data for the internet of things by means of clustering and event processing. In *IEEE International Conference on Communications (ICC)*, 2015, IEEE, pp 685–691.

[18] Naik A: k-Means Clustering Algorithm, 2016, Retrieved June 14, 2016, from https://sites.google.com/site/dataclusteringalgorithms/k-means-clustering-algorithm.

[19] Bishop CM: *Pattern Recognition and Machine Learning,* 2006, Springer.
[20] Monekosso DN, Remagnino P: Data reconciliation in a smart home sensor network, *Expert Syst Appl* 40(8):3248–3255, 2013.

Further reading

[21] Hubel D, Wiesel T: Receptive fields and functional architecture of monkey striate cortex, *J Physiol (Lond)* 195:215–243, 1968.

About the author

Dr. P. Ambika is presently working as an Assistant Professor in Department of Computer Science (PG), at Kristu Jayanti College, Bangalore. She is a Senior IEEE Member and Secretary of IEEE Roof Computing P1931.1 WG and she has over 12 years of experience including academics, research and industry. She has done her PhD in Computer Applications from Anna University in the area of Image Processing and Retrieval in 2014. Dr. Ambika has published more than 12 papers in International/National journals and conferences. She also attended various workshops and FDPs. Her research area includes data science and analytics, AI, machine learning, IoT and roof computing.

CHAPTER FOURTEEN

Energy-efficient edge based real-time healthcare support system

S. Abirami, P. Chitra
Department of Computer Science and Engineering, Thiagarajar College of Engineering, Madurai, India

Contents

Advances in Computers, Volume 117
ISSN 0065-2458
https://doi.org/10.1016/bs.adcom.2019.09.007

Abstract

The ubiquitous usage of wearable IoT (wIoT) devices has created a formidable opportunity for remote health monitoring system to provide paramount services such as preventive care and early intervention for populations at risk. The cloud-edge paradigm can efficiently manifest the complex computations required in providing these services. But the challenge in its exertion lies in incorporating intelligence at the edge devices. With the deluge of data availability, deep learning methods are very promising to obtain sufficient performance in healthcare applications. As the edge devices are resource-constrained in terms of compute capability and energy consumption, unleashing deep learning services from the cloud to the edge requires efficient tackling of the exorbitant computational and energy requirements of deep learning frameworks. In this chapter, an energy-efficient smart edge based health care support system (EESE-HSS) is proposed for diabetic patients with cardiovascular disease. The proposed cloud-edge paradigm makes use of a hierarchical computing architecture for exerting expeditious diagnosis during emergencies. The intelligence framework incorporated at the edge is also built in an energy-efficient manner. Thus, the proposed healthcare support system has better efficacy in terms of energy efficiency and reduced latency. This makes it very supportive for fall detection in diabetic patients with cardiovascular disease who are susceptible to the risk of heart attack, stroke, heart failure and other vicious diseases.

1. Introduction

Health care system provides remote health monitoring for population-at-risk. It performs the task of life saving by preventing worsening of emergency situations for patients under critical health conditions. Cardiovascular disease (CVD) is the major cause of morbidity and mortality for individuals with diabetes. Diabetic patients with cardiovascular disease are prone to complications like coronary heart disease (CHD), stroke, peripheral arterial disease, nephropathy, neuropathy and cardiomyopathy. Cautious health monitoring and fall detection of such patients is indispensable as afterward effects of a fall causes grave effects particularly for aged patients [1]. Preventive measures for such population through active monitoring can be accomplished through the health support system that comes along with monitoring.

Wearable internet of things (wIoT) [2] plays a vital role in healthcare systems for monitoring patients facing health risks. Such IoT-based application extends the boundaries of healthcare outside of hospital settings for early detection and prevention of patient's health deterioration [3]. These applications are intended to provide energy-efficient, low-cost, high convenience and low latency services for healthcare users. Many existing smart health monitoring systems are mostly based on cloud platform [4].

These systems transfers the health related data generated by the IoT devices to the cloud through the Internet and return the diagnostic results obtained through deep learning technology deployed in the cloud. However, such system seems to be inadequate for health care services where low latency is a vital parameter. Thus, the healthcare support system requires a new computational model delay sensitive monitoring with intelligent and reliable control.

Edge computing is a supplement or extension of the cloud computing in which the data is processed near the edge of the network where the data is being generated [5]. Edge computing can also be viewed as a local computing that significantly decreases the consequent traffic, distance and hence the latency. As, the edge devices can respond to data almost instantaneously by its ability to process data without ever putting it into a public cloud, it plays a significant role in latency-sensitive health care applications. In health monitoring system the low-powered wearable sensor devices acts as edge devices. These devices are energy constrained [2] and therefore paradigms and algorithms that reduce the computation and energy consumption without reducing the task performance are indispensable for deploying edge based health care applications.

The cloud-edge computing paradigm is a potential approach for embedding agile computing in health monitoring systems. This model extends to transfer computations between the cloud and edge devices by deploying a hierarchical architecture with the benefits of edge and cloud platforms to assist the analysis of healthcare data from wIoT devices. The incorporation of edge computing makes the model feasible and robust for delay sensitive health care applications while the association of the cloud provides huge resources for computing and storage. Adding intelligence at both cloud and edge can produce vast performance improvement in health care applications.

With the advent of advanced computation platforms, deep learning (DL) is extensively applied for intelligence in many applications such as image classification, object recognition, and natural language processing. The self-taught and compression capabilities of deep learning helps to study the features from the input data hierarchically and automatically and also ensures to illuminate the hidden patterns and unusual patterns from them. This has made deep learning models to be an efficient solution to the most of the IoT-based applications. A wide range of deep learning algorithms are utilized for decision makings in healthcare applications. The efficacy of deep learning is achieved through the deep layers inbuilt in its architecture. This makes the deep learning algorithms computation intensive by nature. Hence the low-powered edge devices are unsuitable for building a deep learning model as it cannot meet the high computational cost demands of deep learning algorithm. The challenge in building an efficient latency aware health monitoring system lies in incorporating the deep learning inference in an edge device that has limited computational capacities [6].

In this chapter, energy-efficient smart edge based healthcare support system (EESE-HSS) is proposed for monitoring diabetic patients with cardiovascular disease. The system observes the patients and discriminates any discrepancy in their health. Alerts and warning to concerned people are sent at critical situations which is a fall due to either a stroke or heart attack. A hierarchical cloud-edge paradigm is utilized which permits energy conservation at the edge device by offloading the high computation task to the cloud. In this, the instantaneous abnormalities are scrutinized by the edge devices to generate alerts. This scrutinization requires only a deep learning model inference and hence the number of computations at the edge is reduced. Through this, the energy expenditure at the edge devices is reduced which in turns enhances the battery life of the wearables. The high computation task such as building and updating models for decision making are made at the cloud which provides the vast computation capacity to build deep learning algorithm with many number of deep layers. This helps to achieve the required system efficiency and adds on to the improved accuracy of the proposed EESE-HSS.

In the proposed EESE-HSS, the deep learning inference is incorporated on the edge devices. Energy efficiency on the inference task can be achieved by associating it with the early exit strategy. Every edge node holds a local hypothesis on it. The latent features of the current input at each layer of the deep network are associated with a confidence value [7]. At every layer the confidence value of the input's latent feature is checked with its associated class labels. If satisfied the process is terminated and if necessary the alert is generated. This early exit during inference helps to conserve energy that could have been exhibited due to the unwanted traversal through the remaining layers [6]. Most apparent situations are designed to hold higher confidence value. Hence the energy required for diagnosing obvious situations is minimized.

The objectives of the proposed work are (i) Monitoring individual diabetic patient with cardiovascular disease using wIoT devices. (ii) Build a cloud-edge paradigm based health support system for diagnosing critical situations for people at risk. (iii) Assimilate intelligence integrated with early exit at the edge devices for energy-efficient smart real-time diagnostics.

2. Edge computing: Principles and architecture

Edge computing is a supplement or extension of the cloud computing that process the data near the data source thereby improving the overall network efficiency and performance with reduced latency [8]. By placing the computations at the edge of the network edge computing lessen the amount of bandwidth required compared to that needed for processing in the cloud. Hence edge computing is a latency aware computing platform that imperatively supports time-critical applications such as healthcare monitoring. The basic architecture of edge computing is as shown in Fig. 1.

The initial part of the network that collects data for processing could either be a IoT network or a Local Area Network or a Radio Access Network the data collected by it are processed locally by the edge devices. After processing they are further transmitted to the cloud for performing high computation tasks and storage purposes.

2.1 Advantages of edge computing over cloud computing

Edge Computing reflects more advantages than the independent cloud computing in the following aspects,

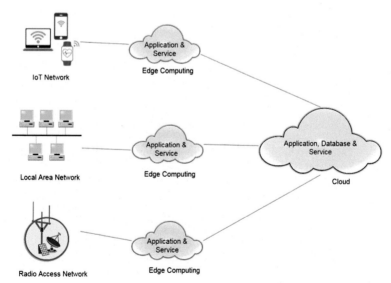

Fig. 1 Basic architecture of edge computing.

- *Efficient bandwidth utilization:* Distributed edge computing nodes can handle large number of computation tasks at the edge device without transmitting the required data to the cloud and hence mitigates the additional transmission pressure of the network.
- *Spontaneous response:* Edge device can handle services. This eliminates the delay of data transmission and improves the response speed.
- *Powerful data storage:* The edge devices gets data back up from the cloud that provides huge processing capabilities and enormous storage.

2.2 Use cases of edge computing

Edge computing facilitates service providers to delve deep into data, perform analytics, gather knowledge and make decisions [8]. It extensively solves the challenges like latency, governance, security and monitoring faced by many application services. Few applications that benefit from edge computing are,

- *Health Care:* The healthcare industry terribly relies on fast-paced services. Even small amounts of latency are undesirable in healthcare application as it can stop patients from accessing vital services. Ensuring responsive healthcare has proved to be a pervasive challenge across the globe. This can be achieved through edge computing that keeps data processing close the source and eliminates the unwanted latency.
- *Smart Cities:* It aims to formulate smarter home via reliable sensors, storage and security [4]. From transportation, buildings to streetlights, everything will be connected and adjoined with intelligence. With edge computing every application would have spontaneous response.
- *5G applications*: The 5G applications are projected to perform faster than 4G applications in terms of responsiveness and bandwidth. Edge computing plays a vital role here in order to make the 5G applications $1000 \times$ faster than 4G.
- *Autonomous vehicles:* Edge computing can help machines to sense, identify and learn things without having to be programmed. This feature of edge computing helps it in building self-driving cars that can process data without the need to connect back to the cloud.
- *Surveillance:* Edge computing provides solutions to video analytics such as facial recognition, event detection and behavioral prediction that helps in security and surveillance. The video surveillance system gathers, process, and analyzes data to detect suspected persons, potentially serious and life-threatening situations and suspicious behaviors. Necessary alerts are then sent to security personal for further action.

- *Software-defined networking:* It requires local processing to determine the best route to send data at each point. Each node and network is equipped to perform decision about the quality of service required for data transmission. This is achieved through edge computing.
- *Blockchain:* Block chain is a distributed ledger technology where every node is a compute unit and hence an edge.

2.3 Limitations of the cloud for healthcare applications

Despite the potential tools and services provided by the cloud technology for IoT applications, it also has various limitations for the implementation of healthcare applications on it. Some of them are

- *Increased latency:* Immediate decisions are imperative in a healthcare application for the need of saving lives. The cloud technology cannot guarantee low latency due to the long distance between client devices and data processing centers.
- *Inefficient utilization of network bandwidth:* Cloud does not endow with unlimited bandwidth to render services regarding the storage and computing facilities available with it.
- *Outage:* Healthcare applications are sensitive applications where disruption of services is undesirable. The cloud technology which is an Internet-based system is vulnerable to technical issues and disruptions in networks.
- *Security and privacy:* Global transfer of data alongside with other user's information in cloud makes it incompatible for healthcare applications that carry sensitive and confidential informations. Cyber attacks or data loss can collapse the objective of the entire health support system.

2.4 Advantages of edge computing for healthcare applications

Edge computing has decentralized cloud architecture [5]. It enables data processing closer to the edge of the network where the data is generated. The following characteristic of edge computing makes it exceptionally appropriate for health care applications,

- *Efficient bandwidth utilization:* Due to data processing at edge devices the amount of data to be transmitted to the cloud for processing is reduced. This in turn reduces the network traffic and improves the system efficiency.

- *Spontaneous Response:* Data processing at the edge devices benefits in producing instantaneous responses to services. This feature of edge computing acts as a life saver in health care monitoring system.
- *Scalability:* Edge computing offers an uncomplicated path to scalability that aids in expansion in the number of patients to be monitored in a health monitoring system. Scalability can be achieved through simply expanding the computing capacity through a combination of IoT devices and edge data centers. The addition of new end user does not impose substantial bandwidth demands on the core of a network due to the use of processing-capable edge computing devices. This feature of edge computing makes it incredibly versatile for health care applications.
- *Reliability:* The processing of the data close to the end user makes edge computing more reliable as it is invulnerable to security threats and network outages.

2.5 Challenges in edge computing deployment

Some of the challenges in offering services through edge computing are

- *Power supply:* The edge devices are energy constrained devices with limited power supply associated with it. The edge has to be able to process at any time instant without any outages. For this, proficient power utilization by the edge devices is mandatory for efficient operation of the system.
- *Deployment:* The edge devices are deficit of space and physical environment. It must be designed to suit the applications environment. The computing capability required at the edge has to be brought within the available space.
- *Maintenance:* Due the distributed architecture proper maintenance is mandatory for proficient working of the system. The device failures have to be sorted beforehand for preventing the disruptions in the system services.
- *Backup:* Data from disparate sources are collected and processed in edge computing. An authentic data protection strategy is crucial for the service providers to comprehend all the collected data. Reliable data storage and access are critical to add security in the application.

3. Deep learning

Deep learning is an eminent subset of machine learning. It utilizes many layers of non-linear processing of information for supervised or unsupervised feature extraction, pattern analysis, classification and prediction. Its efficiency comes through learning multiple levels of useful representation

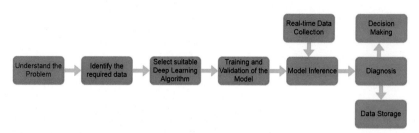

Fig. 2 Deep learning for health monitoring.

and abstraction from data such as images, sound, text and numeric. The most distinctive attribute of deep learning is its efficiency to provide the state-of-the-art accuracy in pattern detection, classification and prediction related tasks. Deep Learning aids in learning more complex features and input-output relationships in a given set of data. Machine learning algorithms require manual feature engineering that has to be explicitly coded by the programmers. Unlike machine learning deep learning algorithms can learn features automatically without any manual inputs.

The overwhelming of data in this big data era is a promising aspect for proficient training of deep learning algorithms. Also, the advent of advanced powerful computing hardware resources such as GPU has an incredible contribution in the success of deep learning over classical machine learning. As an example, Fig. 2 shows the procedures involved for utilizing deep learning in health monitoring.

Some of the popular deep learning models are as follows:

3.1 Multi layer perceptron

Multi layer perceptron (MLP) is a supplement of feed forward neural network. It consists of three types of layers—the input layer, output layer and hidden layer, as shown in Fig. 3. The input layer receives the input signal to be processed. The required task such as prediction and classification is performed by the output layer. An arbitrary number of hidden layers that are placed in between the input and output layer are the true computational engine of the MLP. Similar to a feed forward network in a MLP the data flows in the forward direction from input to output layer. The neurons in the MLP are trained with the back propagation learning algorithm. MLPs are designed to approximate any continuous function and can solve problems which are not linearly separable. The major use cases of MLP are pattern classification, recognition, prediction and approximation.

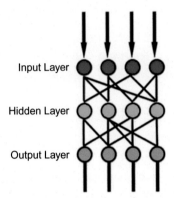

Fig. 3 Schematic representation of a MLP with single hidden layer.

The computations taking place at every neuron in the output and hidden layer are as follows,

$$o(x) = G(b(2) + W(2)h(x)) \tag{1}$$
$$h(x) = \Phi(x) = s(b(1) + W(1)x) \tag{2}$$

with bias vectors $b(1), b(2)$; weight matrices $W(1), W(2)$ and activation functions G and s. The set of parameters to learn is the set $\theta = \{W(1), b(1), W(2), b(2)\}$. Typical choices for s include tanh function with $\tanh(a) = (e^a - e^{-a})/(e^a + e^{-a})$ or the logistic sigmoid function, with $\text{sigmoid}(a) = 1/(1 + e^{-a})$.

3.2 Convolutional neural networks

Convolutional neural networks (CNN) is a biologically-inspired development of MLP. CNN are widely used for image classification, image clustering and object detection in images. They are also employed for optical character recognition and natural language processing. Apart from images, when represented visually as a spectrogram, CNNs can also be applied to sound. Also, CNNs has been applied directly to text analytics as well as in graph data with graph convolutional networks. The state-of-art art efficiency of CNN compared to its baseline algorithms makes it success in many fields.

As, shown in Fig. 4, in CNN the features are detected through the use of filters which are also known as kernels. A filter is just a matrix of values, called weights that are trained to detect specific features. The purpose of the filter is to carry out the convolution operation, which is an element-wise product and sum between two matrices. The training of the CNN is

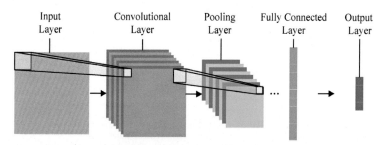

Fig. 4 Schematic representation of convolutional neural networks.

fastened by reducing the amount of redundancy present in the input feature. Hence, the amount of memory consumed by the network is also reduced. One common method to achieve this is max pooling, in which, a window passes over input data and the maximum value within the window is pooled into an output matrix. The algorithm is made efficient for feature extraction by concatenating multiple convolution layers and max pooling operations. The data is processed through these deep layers to produce the feature maps which is finally converted into a feature vector by passing through a MLP. This is referred to as a Fully-Connected Layer that performs high-level reasoning in the developed model.

If the k-th feature map at a given layer is represented as h^k, whose filters are determined by the weights W^k and bias b^k, then the feature map h^k is obtained as follows for tanh activation function

$$h_{ij}^k = \tanh\left(\left(W^k * x\right)_{ij} + b^k\right). \tag{3}$$

The output of the fully-connected layer carries probabilities of each class. The class carrying the highest probability is the classified output. The weights updation and optimization of the algorithm is done through the back propagation of gradients.

3.3 Recurrent neural networks

MLP is incompatible for modeling sequence data. Recurrent neural networks (RNN), a deep learning algorithm, is a promising tool that is proficient in modeling sequence data such as sound, time series sensor data or written natural language. In RNN a feedback loop connects the output from step n-1 to the network and influences the outcome of step n.

This feedback loop differentiates the RNN from the regular feed forward networks. The process of carrying memory forward is mathematically represented as

$$h_t = \Phi(Wx_t + Uh_{t-1}) \tag{4}$$

where s_t is the hidden state at time step t, x_t is the input at the same time step. This is modified by a weight matrix W added to the hidden state of the previous time step s_{t-1} multiplied by its own hidden-state-to-hidden-state matrix U. U is also known as a transition matrix. The extent of importance to accord for both the present input and the past hidden state is determined by the weight matrices. The extension of back propagation known as back propagation through time is utilizes for updating the weights in such a manner that it minimizes the error function. The sum of the weight input and hidden state is squashed by the activation function Φ. Fig. 5 shows the complete sequence of a RNN unfolded into a full network.

The weights in the transition matrix have a strong impact on the learning process of RNN during the gradient back-propagation phase. Small weights in this matrix can lead to a situation where the gradient signal gets too small. This makes the learning process of the model very slow or even stop altogether. This is referred to as vanishing gradients problem. This makes the task of learning long-term dependencies in the input data very tedious. On the other hand, large weights in the transition matrix can lead to large gradient signal causing the learning to diverge. This is known as exploding gradients problem. The problems of vanishing and exploding gradient are eradicated in the LSTM model by the introduction of a new structure called a memory cell that is composed of four main elements: an input gate, a neuron with a self-recurrent connection a forget gate and an output gate as shown in Fig. 6.

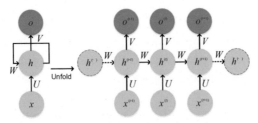

Fig. 5 Schematic representation of recurrent neural networks.

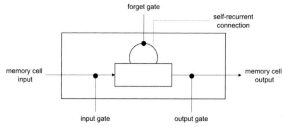

Fig. 6 Structure of a memory cell.

Updation of memory cells at every time step t is described by the equations below:

$$i_t = \sigma(W_i x_t + U_i h_{t-1} + b_i) \tag{5}$$

$$\widetilde{C}_t = \tanh(W_c x_t + U_c h_{t-1} + b_c) \tag{6}$$

$$f_t = \sigma(W_f x_t + U_f h_{t-1} + b_f) \tag{7}$$

$$C_t = i_t * \widetilde{C}_t + f_t * C_{t-1} \tag{8}$$

$$o_t = \sigma(W_o x_t + U_0 h_{t-1} + b_0) \tag{9}$$

$$h_t = o_t * \tanh(C_t) \tag{10}$$

where x_t is the input to the memory cell layer at time t, W_i, W_f, W_c, W_o, U_i, U_f, U_c, and U_o are weight matrices, b_i, b_f, b_c and b_o are bias vectors.

3.4 Restricted Boltzmann machines

Restricted Boltzmann machine (RBM) is an undirected graphical model that falls under deep learning algorithms. It plays an important role in dimensionality reduction, classification and regression. RBM is the basic block of Deep-Belief Networks. It is a shallow, two-layer neural networks. The first layer of the RBM is called the visible or input layer while the second is the hidden layer. The following Fig. 7 shows the model of the RBM. In RBM the interconnections between visible units and hidden units are established using symmetric weights.

From the figure it is seen that each circle represents a node that is connected to each other across layers, but no two nodes of the same layer are linked. Each node is a compute node that processes input and makes stochastic decisions about whether to transmit it or not. Through the several forward and backward passes between the visible layer and hidden layer the RBM learns to reconstruct data by themselves in an unsupervised fashion.

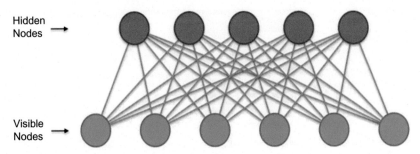

Hidden Nodes ⟶

Visible Nodes ⟶

Fig. 7 Graphical model of the restricted Boltzmann machines.

3.5 Deep-belief networks

A deep-belief network (DBN) is built by appending a stack of RBM layers. In the stack every RBM layer can communicate with both the previous and subsequent layers. Hence it is a network which is assembled out many single-layer networks. Except the first and final layers every layer in a DBN performs a dual role by serving as the hidden layer to the nodes that come before it and as the input layer to the nodes that come after. Some applications of deep-belief networks are to recognize, cluster and generate images, video sequences and motion-capture data. DBN model the joint distribution between observed vector x and the l hidden layers h^k is as follows:

$$P\left(x, h^1, \ldots, h^l\right) = \left(\prod_{k=0}^{l-2} P\left(h^k \mid h^{k+1}\right)\right) P\left(h^{l-1}, h^l\right) \tag{11}$$

where x=ho, $P(h^k \mid h^{k+1})$is a conditional distribution for the visible units conditioned on the hidden units of the RBM at level k, and $P(h^{l-1}, h^l)$is the visible-hidden joint distribution in the top-level RBM. The architecture of a Deep-Belief Network with two RBMs is as shown in Fig. 8.

3.6 Generative adversarial networks

Generative adversarial networks (GAN) has highly potential to mimic any sort of data distribution. This deep learning algorithm has two networks placed opposite to each other. Each network has a different role. One neural network, called the Generator, generates new data instances, while the other, called the Discriminator, evaluates them for authenticity. Generally, the training time of GANs is huge. The generator values are held constant when the discriminator is being trained and similarly the discriminator values are held constant when the generator is trained. Both the generator and

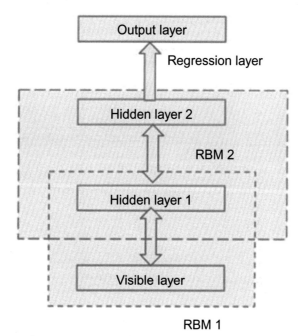

Fig. 8 Architecture of a deep-belief network with two RBMs.

discriminator are trained against a static adversary through a fight between generator and discriminator. Mathematically, it is represented as,

$$\min_{G} \max_{D} V(G, D) \tag{12}$$

$$V(D, G) = E_{x \sim p_{data}(x)}[\log D(x)] + E_{z \sim p_z(z)}[\log (1 - D(G(z)))] \tag{13}$$

It is common for each side of the GAN to overpower the other during training. That is, if the discriminator is too good, the generator will struggle to read the gradient and if the generator is too good, it would persistently exploit weaknesses in the discriminator that lead to false negatives. Fine tuning of their respective learning rates can help to get rid of this issue. The general structure of GAN is as shown in Fig. 9.

3.7 Deep autoencoders

A deep autoencoder is an exceptional deep learning algorithm that constitutes of two symmetrical deep-belief networks with four or five shallow layers, as shown in Fig. 10. Among them half of the network does the

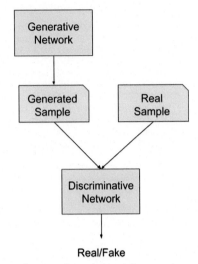

Fig. 9 General structure of generative adversarial networks.

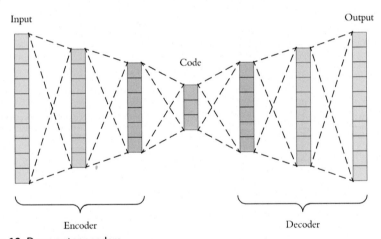

Fig. 10 Deep autoencoders.

job of encoding while the latter half does the decoding. The autoencoders belong to the neural network family. They are also closely related to Principal Components Analysis (PCA) but much more flexible than it. The flexibility with autoencoders is achieved through both linear and non-linear transformation in encoding while PCA can perform only linear transformation. The autoencoders learns significant features present in the data by minimizing the reconstruction error between the input and output data.

In autoencoders, the number of neurons in the output layer is exactly the same as the number of neurons in the input layer. The different types of Autoencoders are

- Under complete Autoencoders
- Sparse Autoencoders
- Denoising Autoencoders (DAE)
- Contractive Autoencoders (CAE)

3.8 Health care application

With the increase in the aging population and rising number of chronic diseases the requirement for health care monitoring system is always high. The advancement of technologies like IoT and cloud had made healthcare easier on a pocket, in terms of accessibility [8–10]. By exploiting the technology-based healthcare method, there are unparalleled benefits which could improve the quality and efficiency of treatments and accordingly improve the health of the patients. Healthcare technology is constantly influenced by innovations in market changes, regulatory changes, and evolving healthcare standards. The most imperative service provided by the heath care systems include real-time monitoring via connected devices to save lives at critical event happenings. On-time alerts are generated during event of life-threatening circumstances. A general network architecture used for health care application is shown in Fig. 11. The system consists of three layers. (1) Collect layer collects the physiological data from the wearable sensor devices. (2) Transmit layer that transfer the collected data to the

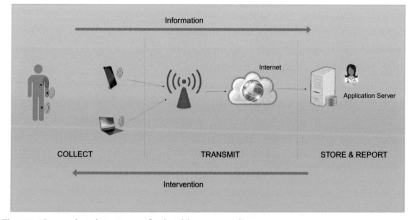

Fig. 11 General architecture of a health care application.

application server via Internet. (3) The application server that performs feature extraction, data analysis and decision making. The advent Artificial Intelligence (AI) also helps people streamline administrative and clinical health care processes [11]. AI rigorously helps frontline clinicians be more productive and in making back-end processes more efficient. It aids in making clinical decisions and improving clinical outcomes.

The proposed system aims to develop a decision support system for diabetes patients with cardio vascular diseases. People having diabetes are more liable to develop heart disease and have greater chance of a heart attack or a stroke. Over exposure to high blood glucose due to diabetes can damage your blood vessels and the nerves that control your heart and blood vessels. Hence the longer one has diabetes, the higher the chances that they would a develop heart disease is greater. In patients with diabetes, the most common causes of death are heart disease and stroke. Patients with diabetes are nearly twice as likely to die from heart disease or stroke as people without diabetes. Efficient health monitoring system for diabetes patients with cardiovascular disease would act as life saving scheme by preventing adverse effects due to critical circumstances.

3.9 Deep learning for health care system

Deep Learning, a layered algorithmic architecture, is a proficient tool for accessing the data collected from wearables and analyzing them for feature extraction and pattern detection. This makes them lend immense service and handle diverse situations in health care applications. Applications of deep learning in health care include diagnosing diseases to drug discovery and manufacturing. Deep learning also helps in predicting epidemic outbreaks for conducting beforehand precautionary measures. Deep learning performs three vital functions in health care applications as shown in Fig. 12. They are anomaly detection, diagnosis and prognosis.

In health care applications, anomaly detection is an essential task that is crucial for improving the quality of health services and avoids loss of lives. Anomaly detection performs the task of identifying any irregularities in the treatment and thereby prevents situations that are extremely dangerous, and even deadly. It also detects the potentially fraudulent activities which in turn help to recover costs and lessen the overall impact of fraud on the healthcare system. Diagnosis refers to spotting a disease or a condition with respect to a patient's symptoms and signs. An outcome of diagnosis relies on the patient's previous history. Prognosis is a prediction or estimate of the chance of

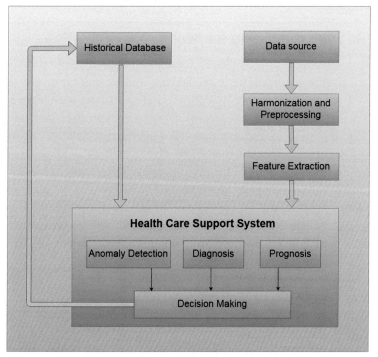

Fig. 12 Role of deep learning in health care application.

recovery or survival from a particular disease. Most prognosis is based on statistics of how a disease acts in studies on the general population. Based on the signs and symptoms, prognosis intimates about length of the disease existence; expectations of quality of life, such as the ability to carry out daily activities; the potential for disease complications and rise of associated health issues; and the likelihood of survival.

Deep learning which is a type of machine learning is inspired by the human brain. It consists of deep layers of neural network architectures and they learn from large amounts of data. Deep learning algorithms eliminate the need for feature extraction manually as they automatically learn features, thus avoiding a large amount of time-consuming engineering. These features of deep learning make it very appropriate for performing the tasks of anomaly detection, diagnosis, and prognosis in health care support system. The deep architecture of deep models makes it consume more energy for providing accuracy. This can be overcome through the usage of adaptive deep learning algorithms [12].

4. Edge computing for health care application

Edge computing is employed for a wide range of applications [13]. Unlike any other applications, the healthcare applications rely on fast-paced service. Even small amounts of latency are undesirable in a healthcare application as it can hinder patients from accessing vital services [14]. Hence, ensuring responsive healthcare remains a pervasive challenge across the globe. Edge computing overcomes this by providing increased performance and responsiveness in healthcare applications. Reduced latency in edge computing is accomplished with data processing at the edge of the network rather than the center. Today, healthcare providers are looking for the most effective ways to maximize the healthcare coverage. Wearable IoT devices diagnosing or monitoring conditions will reduce the number doctor appointments required [3].

In edge computing based health care system the deployed IoT devices gather and analyze data from patients locally. The wearable IoT devices collect physiological data from the patients and transmit them to the application servers [14]. Edge computing pushes the computing and processing capabilities closer to where the data originates [8,12]. This helps to access data when internet connections are deficient for access of centralized databases. Thus, embracing edge computing for health care system can give more control and the ability to leverage data in near real-time, which remains a promising strategy to help service providers distribute their network and data in such a way that it can be used and shared quickly and securely.

Managing the energy efficiency on a energy limited edge device is the most challenging part in designing a service architecture through edge computing [12,15,16]. Smart edge devices that overcome this challenge paves path for smart and sensitive applications. Smart healthcare is one of the major components of smart cities [4]. The field of smart healthcare emerges from the need to improve the management of healthcare sector, better utilize its resources, and reduce its cost while maintaining or even enhancing its quality level. The need for scalable healthcare services to patient with real-time and cost-effective patient remote monitoring stimulates the necessity of smart edge that performs feature extraction, at edge devices for event detection [11,17]. This aid in instantaneous detection of critical situations without any delay, thereby saving lives on a higher extend. The smart edge based health care monitoring system is illustrated in Fig. 13.

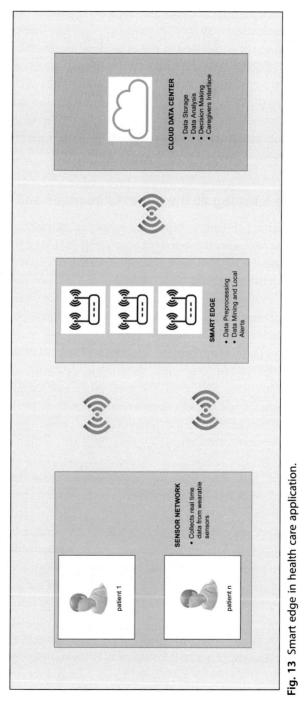

Fig. 13 Smart edge in health care application.

Initially the sensor nodes collect real-time data from the wearable sensors worn by the patients. After data preprocessing, the immediate inference for critical event is inferred at the edge device with the help of the detection model imparted on the edge devices. Immediate emergency messages and alerts are generated in case of critical events. The collected data are further moved to the cloud platform. With the enormous computing capability available with it the cloud platform takes care of data analyzing for decision making, data storage and communicating the care givers in case of emergencies.

5. Deep learning at the edge: Challenges and benefits

Deep Learning and edge computing are two potential two that individually can revolutionize the health care sector [17]. Consolidating both together is intriguing as it brings out new opportunities in enabling ideal health care services to the population that need it the most. Deep learning at the edge devices invites many challenges along with its benefits. Deep learning at the devices enjoys the twin benefits of reduced latency in services and reduced cost of data transmission for processing. Smart edge dramatically reduces the amount of data that has to be sent over the network, thereby reducing network congestion, speeding up operation and, in many instances, reducing costs [18]. Also, the performance of the application is typically much better, because of the processing been done at the edge itself rather the cloud. Smart edge helps to maintain the privacy of the patients and makes the system more reliable by making it susceptible to network outages [18].

Though intelligent edge enjoys numerous benefits many challenges also remains coupled with it [17]. Among them the most vital challenge to be taken care is the realization of energy efficiency in the deployed algorithms [8]. The edge devices utilized in health care are the wearable IoTs that are compatible low power devices. Incorporating computing facility on them require plenty of challenges to be dealt beforehand. Energy-efficient algorithms on the edge minimize its power consumption and increases the battery time of the edge device [15,19]. Deep learning algorithms are deep layered architecture that mimics human brain. More number of neuron refers to the increase in power dissipated for computation at the edge.

Deep learning incorporates more number of neurons in its deep architecture. The energy consumed for the computations is high. Incorporating such an energy demanding algorithm on energy constrained edge device for the sake of achieving the state-of-art accuracy demands many challenges to be solved [20]. The deep learning architecture for training and validation has to be made energy efficient in order to make its implementation efficient on the edge devices. Speeding up deep learning inference and performing distributed training on edge devices are two possible solutions for incorporating deep learning at edge devices.

6. Proposed system

In this, an energy-efficient smart edge based health care support system (EESE-HSS) is proposed for remote health monitoring of diabetes patients with cardio vascular diseases. High glucose levels for a long time period could lead to stroke and heart attacks. Hence the diabetes patients with cardio vascular disease are highly susceptible to sudden strokes and heart attacks. The proposed EESE-HSS aims in providing instantaneous response to such patients under critical events. For, old age patients the immediate response over a fall due to stroke or heart attack is mandatory before any therapies for the disease. The fall is considered as a critical event for which alerts has to be generated to the concerned people with not much delay. This necessitates the incorporation of the deep learning model for fall detection at the edge for spontaneous response. Further the data are transmitted to the cloud for advanced diagnosis of the disease. Also, the collected data at the sensor networks are stored in the cloud for future references and online updation of the DL model.

Deep learning based fall detection using the physiological data collected by the wIoT is built in the cloud and a local hypothesis of it is incorporated on the edge. Inference to this aids fall detection at the edge. If the classified output at the edge inference is a fall then immediate alerts are generated to emergency wards. This helps to prevent adverse effects in patients particularly at old age. The architecture of the proposed EESE-HSS based on smart edge is as shown in Fig. 14.

The proposed energy-efficient smart edge based health care support system (EESE-HSS) is a three-tier architecture encompassing deep learning and intelligent edge. The data collected from the wearable IoT devices are passed through three layers for efficient remote health monitoring of diabetes patients with cardiovascular disease. The responsibilities of each layer is as follows:

- *WPAN:* This layer is composed of different types of wearable IoTs that helps to collect multimodal physiological data from patients that aid in ideal diagnosis during health care monitoring. These data are collected

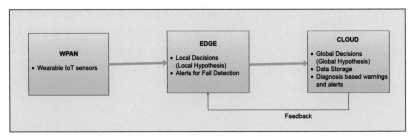

Fig. 14 Architecture of the proposed energy-efficient smart edge based health care support system.

and harmonized for transmission to the edge device. Sometimes the edge device is the wearable IoT itself.

- *Edge:* The edge utilized in EESE-HSS is an intelligent edge that has the deep learning based model for fall detection. The edge device holds a local hypothesis for detecting a fall through the data collected by the wearable IoTs. In case of positive classification immediate alerts are sent to concerned family members or hospitals. The edge device here holds only the inference part of the deep learning model. Based on the streaming input data the model for fall detection is updated at the cloud and its updates are fed back to the edge in order to update the local hypothesis associated with it.

- *Cloud:* The massive computing capability of the cloud is utilized for online building of the deep learning model for diagnosis and decision making. The global hypothesis built on the cloud is exploited for disease detection. The cloud also enables every patients timely updated clinical records and the actions taken at the edge and cloud as a responsive measure. Through the global hypothesis the local hypothesis at the edge is periodically updated in order to improvise the fall detection with respect to the streaming inputs collected through the sensors.

The proposed EESE-HSS is made energy efficient by incorporating the cloud-edge paradigm and early exit at inference at the edge. The energy consumed at the edge is reduced thereby extending the battery life of the energy constrained edge devices. This in turn reduces in the outages possible in the application service.

7. Cloud-edge paradigm

The proposed energy-efficient smart edge based health care support system (EESE-HSS) employs a cloud-edge paradigm as shown in Fig. 15 that frames the heath care monitoring application over a distributed system

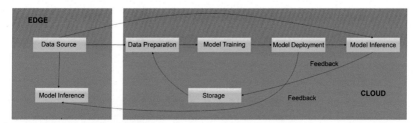

Fig. 15 Cloud-edge paradigm.

abstraction enabling the deployment of fall detection on the energy constrained edge platforms, coupled with more complex diagnosis and decision making on the large scale datacenters (cloud). This paradigm is driven by the significant decrease in energy consumption at the edge along with data storage support by the cloud. The reduction in the energy consumption at the edge is obtained by moving the complex computations through the cloud while maintaining only the crucial ones at the edge. Synchronization between the cloud and edge is periodically established through feedbacks in order to update the local hypothesis present at the edge devices through the global hypothesis that was built by training the deep learning model through the online sequential input data transmitted from the sensor devices. At every instant the collected data is stored and processed for the next updation of the global hypothesis present in the cloud. The updated model is deployed in the cloud for diagnosis and decision making. The updated model is also imparted at the edge device for improving the accuracy in fall detection with respect to the streaming input data.

8. Early exit at inference

While deeper deep learning algorithms achieve maximum accuracy, for many inputs shallower networks are sufficient [7]. With reference to this, not all layers of the deep learning architectures incorporated at the edge have to be traversed before inferring fall detection. Easy data points as shown in Fig. 16 that lie far from the decision boundary can be inferred with very less layers itself. After every layer a linear classifier is placed to produce the confidence value associated with classified latent variables at that particular layer. If a high confidence value is associated with the class fall is greater than a threshold value say 'T,' then the data is classified as fall. Immediate alerts to the concerned people are generated and the data is further transmitted to the cloud for diagnosing the type of disease say a stroke or heart attack.

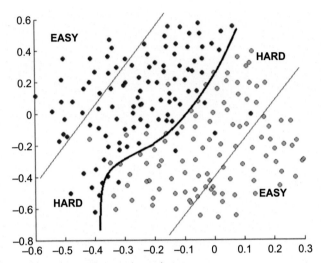

Fig. 16 Easy and hard classification boundaries.

If a high confidence value is associated with the class not a fall is greater than a threshold value say 'T' then the data is carried to the cloud for storage and model updation purpose.

Early termination during inference for easy inputs reduces the energy consumption that would have occurred due to the unwanted traversal of layers in the deep architecture of deep learning algorithms. For this adaptive layer utilization the confidence value is calculated using the entropy value given as.

$$entropy\,(y) = \sum_{c\varepsilon C} y_c \log y_c \tag{14}$$

where y_c represent the probabilities of output class labels while C represents all the class labels available. If the value of y_c is greater than T then it is confident about the output class and it can immediately exit or else it has to move on to the next layer. Therefore the easy inputs are classified within first few layers while the hard inputs have to move deeper. This is illustrated in Fig. 17.

8.1 Hardware and software required for realization of the proposed system

The need for energy-efficient smart edge based health care support system in the proposed architecture can be fulfilled only when the implemented edge

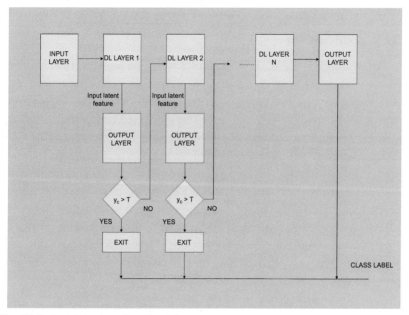

Fig. 17 Early exit in deep learning inference.

devices are proficient to perform deep learning inferences at low energy consumption. A wide variety of edge nodes such as Raspberry Pi or the embedded microprocessor Intel Edison, wearable IoTs with various hardware and software capabilities can be chosen for the proposed application. The required deep learning model for inference can be built through any high-level programming languages (e.g., C++, Python, and Java.) using any real-time database available for fall detection (e.g., Fall Detection Data from China in Kaggle-Activity of elderly patients along with their medical information) on every edge node. Similarly the deep learning based and disease diagnosis model can be developed through any high-level programming languages (e.g., C++, Python, and Java) in the cloud using the real-time database available for disease prediction (e.g., Heart Disease and Stroke Prevention dataset provided by the National Cardiovascular Disease Surveillance System). Proper communication link via gateways and are established between the cloud and edge layer. This facilitates the transfer of data that have been aggregated and computed at the edge layer to the cloud and the updates the local hypothesis on the edge nodes through the global hypothesis that are created based on the deeper analytics in the cloud.

9. Conclusion

The proposed EESE-HSS enables a real-time smart health diagnosis with an energy-efficient cloud-edge paradigm and early exit inference. The proposed system achieves significant energy reduction without any compromise in accuracy of the system. The employed cloud-edge paradigm distributes the function of critical event detection-fall detection at the edge and the disease prediction at the loud. This helps to generate instantaneous alerts at critical situation without any latency in the service. Dynamic online model training at the cloud predicts disease with better accuracy for better decision making. Distribution of application function among edge and cloud platform along with the early exit of inference at the edge device helps in mitigating the superfluous energy consumed in providing health care services. Hence the proposed architecture EESE-HSS provides remarkable heath care services with minimized energy dissipation. This makes it very significant for providing remote health care monitoring for diabetes patients with cardiovascular disease for whom spontaneous response services to stroke and heart attack like critical events are mandatory for saving lives.

References

[1] Queralta JP, Gia TN, Tenhunen H, Westerlund T: Edge-AI in LoRa-based health monitoring: fall detection system with fog computing and LSTM recurrent neural networks. In 2019 42nd International Conference on Telecommunications and Signal Processing (TSP). 2019, IEEE, pp 601–604.
[2] Soundarabai PB, Chelliah PR: Networking topologies and communication technologies for the IoT era. In *Connected Environments for the Internet of Things*, Cham, 2017, Springer, pp 241–268.
[3] Azimi I, Takalo-Mattila J, Anzanpour A, Rahmani AM, Soininen JP, Liljeberg P: Empowering healthcare IoT systems with hierarchical edge-based deep learning. In 2018 IEEE/ACM International Conference on Connected Health: Applications, Systems and Engineering Technologies (CHASE). 2018, IEEE, pp 63–68.
[4] Janet B, Raj P: Smart city applications: the smart leverage of the internet of things (IoT) paradigm. In *Novel Practices and Trends in Grid and Cloud Computing*, 2019, IGI Global, pp 274–305.
[5] Raj P, Pushpa J: Expounding the edge/fog computing infrastructures for data science. In *Handbook of Research on Cloud and Fog Computing Infrastructures for Data Science*, 2018, IGI Global, pp 1–32.
[6] Parsa M, Panda P, Sen S, Roy K: Staged inference using conditional deep learning for energy efficient real-time smart diagnosis. In 2017 39th Annual International Conference of the IEEE Engineering in Medicine and Biology Society (EMBC). 2017, IEEE, pp 78–81.
[7] Panda P, Sengupta A, Roy K: Conditional deep learning for energy-efficient and enhanced pattern recognition. In 2016 Design, Automation & Test in Europe Conference & Exhibition (DATE). 2016, IEEE, pp 475–480.
[8] Ko JH, Na T, Amir MF, Mukhopadhyay S: Edge-host partitioning of deep neural networks with feature space encoding for resource-constrained internet-of-things platforms. In 2018 15th IEEE International Conference on Advanced Video and Signal Based Surveillance (AVSS). 2018, IEEE, pp 1–6.

[9] Evangeline DP, Anandhakumar P: Non-rigid registration of brain MR images for image guided neurosurgery using cloud computing. In *Computer Aided Intervention and Diagnostics in Clinical and Medical Images*, Cham, 2019, Springer, pp 49–59.

[10] Dhanya NM, Kousalya G, Balarksihnan P, Raj P: Fuzzy-logic-based decision engine for offloading IoT application using fog computing. In *Handbook of Research on Cloud and Fog Computing Infrastructures for Data Science*, 2018, IGI Global, pp 175–194.

[11] Hao Y, Jiang Y, Hossain MS, Alhamid MF, Amin SU: Learning for smart edge: cognitive learning-based computation offloading. In *Mobile Networks and Applications*, 2018, Springer, pp 1–7.

[12] Stamoulis D, Chin TWR, Prakash AK, Fang H, Sajja S, Bognar M, Marculescu D: Designing adaptive neural networks for energy-constrained image classification. In Proceedings of the International Conference on Computer-Aided Design. 2018, ACM, p 23.

[13] Khan S, Muhammad K, Mumtaz S, Baik SW, de Albuquerque VHC: Energy-efficient deep CNN for smoke detection in foggy IoT environment, *IEEE Internet Things J* 2:1, 2019.

[14] Devarajan M, Subramaniyaswamy V, Vijayakumar V, Ravi L: Fog-assisted personalized healthcare-support system for remote patients with diabetes, *J Ambient Intell Humaniz Comput*: 10:3747–3760, 2019.

[15] Yang, K., Shi, Y., Yu, W., & Ding, Z: Energy-Efficient Processing and Robust Wireless Cooperative Transmission for Edge Inference. arXiv preprint arXiv:1907. 12475. vol. 1, 2019, pp 1–13.

[16] Zhang Q, Lin M, Yang LT, Chen Z, Khan SU, Li P: A double deep Q-learning model for energy-efficient edge scheduling, *IEEE Trans Serv Comput*, 12(5):739–749, 2018.

[17] Han, Y., Wang, X., Leung, V., Niyato, D., Yan, X., & Chen, X. (2019). Convergence of Edge Computing and Deep Learning: A Comprehensive Survey. arXiv preprint arXiv:1907.08349. vol. 1, pp 1–29.

[18] Yu J, Fu B, Cao A, He Z, Wu D: EdgeCNN: a hybrid architecture for agile learning of healthcare data from IoT devices. In 2018 IEEE 24th International Conference on Parallel and Distributed Systems (ICPADS). 2018, IEEE, pp 852–859.

[19] Toor A, ul Islam S, Ahmed G, Jabbar S, Khalid S, Sharif AM: Energy efficient edge-of-things, *EURASIP J Wirel Commun Netw* 2019(1):82, 2019.

[20] Roy D, Srinivasan G, Panda P, et al: Neural networks at the edge. In 2019 IEEE International Conference on Smart Computing (SMARTCOMP). 2019, IEEE, pp 45–50.

About the authors

S. Abirami is currently a research scholar in Department of Computer Science & Engineering, Thiagarajar College of Engineering. She completed her B.E from National Engineering College, Kovilpatti. She completed her M.E from Anna University Regional Campus, Coimbatore. Her areas of interest includes spatiotemporal modeling, deep learning, and cloud computing.

Dr. P. Chitra is currently working as Professor in Department of Computer Science & Engineering, Thiagarajar College of Engineering. She completed her B.E from Madurai Kamaraj University during 1995; subsequently she worked as lecturer and completed her M.E and Ph.D in CSE during 2004 and 2011, respectively. She is a reviewer for many national and international peer reviewed journals and member of technical program committee for many IEEE national and international conferences. She has under her credits many publications in reputed international conferences and journals in the areas of distributed systems, cloud computing, Multicore architectures.

Printed in the United States
by Baker & Taylor Publisher Services